Drug Addiction Mechanisms in the Brain

Authored by

Jayalakshmi Krishnan

Department of Life Sciences
Central University of Tamil Nadu
Tamil Nadu
Thiruarvur, India

Drug Addiction Mechanisms in the Brain

Author: Jayalakshmi Krishnan

ISBN (Online): 978-981-5223-82-8

ISBN (Print): 978-981-5223-83-5

ISBN (Paperback): 978-981-5223-84-2

need for a court order if at any point you breach any terms of this License Agreement. In no event will any delay or failure by Bentham Science Publishers in enforcing your compliance with this License Agreement constitute a waiver of any of its rights.

3. You acknowledge that you have read this License Agreement, and agree to be bound by its terms and conditions. To the extent that any other terms and conditions presented on any website of Bentham Science Publishers conflict with, or are inconsistent with, the terms and conditions set out in this License Agreement, you acknowledge that the terms and conditions set out in this License Agreement shall prevail.

Bentham Science Publishers Pte. Ltd.
80 Robinson Road #02-00
Singapore 068898
Singapore
Email: subscriptions@benthamscience.net

BENTHAM SCIENCE

CONTENTS

PREFACE

It is generally believed that drug abuse can cause severe long-lasting changes in the neural network contributing to the development of addiction. Profound states of addiction may be established in brains with repetitive usage, despite its damage to the brain. Dopamine plays a very significant role in addiction to drugs. Under normal conditions, in neural communication between neurons, the presynaptic neuron releases dopamine into the synapse. At the postsynaptic neurons, there are receptors which receive the dopamine. Usually, any left-out dopamine molecule is recycled back to the presynaptic neuron by the dopamine transporters. If any drug blocks the dopamine transporter it is unable to take the dopamine back from the synaptic cleft leading to the continuous firing of neurons. As per Drug Enforcement Administration (DEA) and the Controlled Substances Act (CSA) reports, heroin is a Schedule I drug. Heroin causes addiction to the brain like any other addictive substance. Heroine use affects not only neurotransmitters but also the hormonal systems in an irreversible way. In healthy young people, the use of MDMA can lead to cognitive decline when abused with cannabis. MDMA causes hyponatremia and hyponatremia-associated deaths. This book deals with the harmful effects of drugs on brain and cognitive functions. I wish our readers can be satisfied with many questions and feel excited to find the answer to the research questions on the etiology of neurological sequelae of drug abuse in this book.

Jayalakshmi Krishnan
Department of Life Sciences
Central University of Tamil Nadu
Tamil Nadu
Thiruarvur, India

CHAPTER 1

Cocaine and its Effects on the Brain

Abstract: Brain's limbic system is the target site of action of cocaine. This area of the brain is involved in pleasure and motivation. Cocaine causes the dopamine build-up in the synapses by creating a feeling of being "high". Cocaine induces action by binding to the dopamine transporter, which transports excess dopamine back to the presynaptic neuron. The nucleus accumbens (NAc) of the limbic system is the primary target of cocaine action. Cocaine also alters gene expression in the limbic system by altering dopamine transporters or dopamine receptors. Cocaine causes auditory hallucinations, restlessness, paranoia, and psychosis. This chapter reviews the impact of cocaine on the brain.

Keywords: Cocaine, Limbic system, Nucleus accumbens (NAc).

INTRODUCTION

It is generally believed that drugs of abuse can cause severe long-lasting changes in the neural network contributing to the development of addiction. Cocaine is the most powerful reinforcing drug of abuse which can bind to serotonin transporter (SERT), dopamine transporter (DAT), and norepinephrine transporter (NET) by causing blocking of the reuptake of these neurotransmitters. Cocaine is a CNS stimulant that alters sleep and causes alertness. Withdrawal of cocaine is always accompanied by lack of motivation, increased irritability, agitation, extreme fatigue, and depression. Anxiety is one of the main symptoms of cocaine withdrawal and corticotropin releasing factor/hormone (CRF or CRH) has been involved in cocaine abstinence.

Cocaine is synthesized from the coca plant, a native of South America. It is sold as a solid rock crystal form or a fine white powder. Cocaine can be snorted, rubbed in water in the gums, and can be injected with a needle. Another method of taking cocaine is just heating up the rock crystal and directly inhale.

Profound states of addiction to cocaine may be established in brains with repetitive usage, despite its damage to the brain. Studies in animal models prove that the limbic system, basal ganglia, and ventral striatum are involved in dopaminergic neurons that cause pleasure experience [1 - 3]. Continuous use of cocaine can lead to damage to the structural components of the brain, causing mental health disorders such as anxiety and depression and loss of gray matter in the brain, followed by the death of neuronal cells. Cocaine also damages breathing, the immune system, heart dysfunctions, and digestive system problems. Cocaine alters the signalling of neurotransmitters in the brain. Information processing and emotions are affected by cocaine addiction, and also the prefrontal cortex becomes sensitive to cocaine addiction. Although some of the effects of cocaine on the brain are known, many things remain unknown, especially the chronic changes associated with cocaine exposure. Cocaine exposure also alters the learning process within the striatum and prefrontal cortex according to some animal studies. Low doses of cocaine can be dangerous. A single low dose of cocaine can cause structural brain damage in Balb-c mice (o.5mg/kg) without altering the metabolism [4]. As per the European drug report, 4.3 million people between 15 and 64 years old have used cocaine (European Drug Report, 2016). Cocaine also causes myocardial infarction and psychiatric illness [5].

In vitro and *in vivo* experiments have shown a different kind of trend when cocaine is administered. Cocaine causes less firing in neurons *in vivo*, whereas deep hyperpolarization *in vitro*. Withdrawal of cocaine leads to the impairment of sodium currents in nucleus Accumbens neurons [6]. Not only sodium currents but calcium homeostasis is also affected by cocaine in the nucleus Accumbens neurons [7]. Calcineurin is a calmodulin-dependent serine/threonine protein phosphate whose levels were decreased in neurons due to cocaine uptake (Hu *et al.*, 2005). Whole cell calcium levels are affected due to chronic cocaine intake and distups, synaptic plasticity, and intracellular signaling cascades with substantial changes in neurotransmitter release [8].

Cocaine in Brain

Under normal conditions, in neural communication between neurons, the presynaptic neuron releases dopamine into the synapse. In the postsynaptic neurons, there are receptors that receive this dopamine. Usually, any left-out dopamine molecule is recycled back to the presynaptic neuron by the dopamine transporters. Cocaine intake binds with the dopamine transporter and blocks the normal recycling process. This results in the build up of dopamine in the synapses, contributing to the pleasurable effects of cocaine. Cocaine is also a psychostimulant [9]. In brain, cocaine usage can create a short-term change such as alertness, feeling of pleasure, increased energy, overactive, paranoia, *etc.* Significant neurological adaptation can be seen in mice after cocaine exposure in terms of the release of the excitatory neurotransmitter glutamate [10]. Cocaine use is also related to stress and both co-occur at any time [11]. Stress pathway and reward pathway are different in brain, but both can overlap by connections from the ventral tegmental area. Functioning of the orbitofrontal cortex (OFC) is also reduced due to chronic cocaine exposure leading to poor decision making [12]. Ventral pallidum (VP) is connected to the nucleus Accumbens *via* both direct and indirect pathways. Cocaine reinforcement is mediated through Ventral pallidum (VP), which is a part of the basal ganglia. Cocaine inhibits the indirect pathway of neuronal synaptic transmission in VP [13]. By inhibiting the reuptake of 5-HT, cocaine increases extracellular concentrations of serotonin (5-hydroxytryptamine or 5-HT) in the nucleus accumbens [NAc] and ventral pallidum. Cocaine also disturbs the learning and memory pathways of brain. Long-term potentiation is affected due to cocaine administration. Transcription factor ΔFosB in longterm cocaine administration, causes NMDR activation in NAC [14]. ΔFosB concentrations are increased in reinforcing effects of cocaine.

Effects of Cocaine on Brain Cells

Cocaine releases CXCL10 from pericytes and it regulates monocyte transmigration into the CNS [15]. Cerebrovascular accidents are the most common form of cocaine abuse [16]. The neurotoxicity of cocaine addiction is also due to oxidative stress, autooxidation, and apoptosis [17, 18]. Continuous exposure to a psychostimulant drug can lead to changes in cerebral glucose metabolosim [19, 20]. Cocaine causes autophagic cytotoxicity by activating the nitric oxide GAPDh signaling cascade [21]. Mesocorticolimbic dopamine system is the main reason for cocaine seeking behaviour when activated [22]. Cocaine and methamphetamine users experience changes in the orbitofrontal cortex. OFC and medial prefrontal cortex (mPFC) in rats, when analysed for spine density, revealed a profound change [23]. Alertness, attention, and energy are elevated in cocaine users.. In hippocampal neurons and astrocytes glial fibrillary acidic protein is expressed in response to cocaine administration [24]. When BV2 microglial cells were exposed to cocaine, it altered exosome biogenesis [25].

Cocaine at the Synapse

Cocaine affects DA levels in the mesolimbic reward pathway. Cocaine binds to the dopamine transporter (DAT), hence blocking the reuptake of dopamine in the presynaptic terminal. Because of this, the extracellular dopamine level is increased by much magnitude. DAT is a protein located in the presynaptic neuron, and the functioning of DAT is essential for proper dopamine neurotransmission [26].

Cocaine can also increase all monoamine neurotransmitter levels in the brain, not only dopamine. Serotonin and norepinephrine levels are also increased by cocaine use. The reinforcing effects of cocaine are due to the dopamine levels. Cocaine also alters NMDA dependent signal transduction in striatal neurons. Repeated cocaine exposure decreases NMDAR interactions with the postsynaptic density (PSD), and synaptic lipid rafts in the accumbens shell and dorsal striatum in mice [27]. Sigma-1 and Sigma-2 Receptor expression modulate the effect of cocaine on dopaminergic transmission [28]. D1 medium spiny neuron subtype operates in a subtype- and projection-specific manner to negatively regulate cocaine addiction [29]. Remodelling of the mesocorticolimbic circuitry is the main reason for drug addictive behaviour in cocaine users.

Impaired function of Ca^{2+}-activated small-conductance calcium-dependent potassium (SK) channels increases the firing of dopamine neurons in the ventral tegmental area (VTA) dopamine (DA) neurons. A single injection of cocaine within hours induces a lot of changes in the synaptic strength of excitatory inputs in the ventral tegmental area [30, 31]. Changes in AMPAR and NMDAR subunit composition lead to long-lasting neural strength that causes addictive behavior. This drug-evoked synaptic plasticity acts as an intensive signal to promote drug seeking behaviour [32]. Calcium-Impermeable NMDARs mediate cocaine induced excitatory activity in the ventral tegmental area [33]. There is evidence that various classes of drugs of abuse can strengthen the synapse, especially the excitatory synapses on the midbrain region of DA neurons. Cocaine- and stress-induced synaptic enhancement is due to the upregulation of α-amino-3-hydro-y-5-methyl-4-isoxazolepropionic acid receptor subunit GluRA [34]. Excitatory synaptic transmission within the VTA region contributes to behavioural plasticity [35]. GABA synaptic transmission is also disturbed in neurons that project into the ventral pallidum. There is no treatment for cocaine addiction currently. Cognitive behavioural therapy is the only method of treatment. Any specific change in normal synaptic neurotransmission is said to be synaptic plasticity. Molecular targets are the impending features for cocaine to induce long-term synaptic changes. Drug-evoked synaptic plasticity causes changes in the mesolimbic pathways, and thus altering the functional reward circuitry [36]. It is known that cocaine infusion in rats leads to the downregulation of or suppression of the activity of dopamine and serotonin neurons [37].

Cocaine and Genetic Changes in Brain

Epigenetic changes cause both heritable and stable changes in gene expression without much altering the DNA sequence. DNA in mouse brains is altered due to the use of cocaine in the mouse brain, especially those areas which are involved in reward systems [38]. Epigenetic changes have occurred in mouse brains after cocaine intake, followed by changes in the types of RNAs cells made through splicing. Many kinds of proteins are made due to the epigenetic changes in the mouse brain. Gene expression is enhanced upon repeated cocaine exposure leading to addictive phenotypes [39]. Even cocaine induced epigenetic changes are inherited in the germline passing to the offspring. Such studies may shed light on the epigenetic mechanisms that lead to heritable changes in cocaine addiction. Enzymatic modifications to the DNA sequence can occur due to epigenetic changes after cocaine exposure. DNA methylation is one of the modifications of DNA sequence. Cocaine-induced behavioral responses are noted in conditions when cocaine is able to do changes in DNA methylation. Acute cocaine exposure upregulates DNA methylation, DNMT3A and DNMT3B levels leading to the downregulation of gene expression in NAC [40]. Cocaine causes structural plasticity due to DNA methylation. Dendritic spine density is increased in cocaine exposure due to the upregulation of DNMT3a [41]. Histone acetylation is another example of genetic modification. Acute cocaine administration increases histone H4 acetylation on immediate early genes *Fos* and *Fosb* in NAC area [42, 43]. In the striatum, histone H3 phospho-acetylation and histone H4 acetylation increase due to acute cocaine exposure [44]. In cocaine-addicted mouse brain, aquaporins have deletions and have shown to decrease dopamine levels and glutamate levels as well in the nucleus Accumbens region [45]. GDNF (Glail cell-derived neurotrophic factor) is decreased due to the decreased level of protein kinase that is phosphoRet, causing behavioural sensitivity to cocaine [46]. Cocaine induces fetal brain retardation due to the inhibition of macromolecular synthesis [47]. Astrocytes cells express corticotropin-releasing factor (CRF) in the ventral tegmental area due to cocaine exposure [48]. JAK/STAT pathways mediates the effect of chronic administration of cocaine in dopaminergic neurons [49]. Drug (Cocaine) induced adaptations in the rat brain are caused by increased phosphorylation of ERK in VTA, thus, in turn, regulates the increase in tyrosine hydroxylase, a crucial enzyme in dopamine synthesis [49]. Decreased neurofilament levels upon exposure to cocaine are shown to be the reason for the structural alterations seen in the brains of rats [50]. Region-specific effects of cocaine are also observed in the brain. For example, in Nucleus accumbans, cocaine decreased tyrosine hydorxylase phosphorylation, and did not alter the functions of this enzyme in other brain regions such as caudoputamen, and substantia nigra [51]. A single dose of cocaine (0.5 mg/kg) in Balb-c mice has been shown to cause structural brain injury [52]. In adult squirrel monkeys, cocaine administration and imaging by fMRI have revealed an accumulation of cocaine in the dorsal anterior cingulate (dACC) region of the brain [53 - 57].

CONCLUSION

Cocaine is the most powerful reinforcing drug of abuse which can bind to the serotonin transporter (SERT), dopamine transporter (DAT), and norepinephrine transporter (NET) by causing blocking of the reuptake of these neurotransmitters. Cocaine is a CNS stimulant that alters sleep and causes alertness. Cocaine intake binds with the dopamine transporter and blocks the normal recycling process.

This results in the build up of dopamine in the synapses, contributing to the pleasurable effects of cocaine. Cocaine is also a psychostimulant. The impaired function of Ca^{2+}-activated small-conductance calcium-dependent potassium (SK) channels increases the firing of dopamine neurons in the ventral tegmental area (VTA) dopamine (DA) neurons. Cocaine causes structural plasticity due to DNA methylation. Dendritic spine density is increased in cocaine exposure due to the upregulation of DNMT3a.

REFERENCES

[1]   Louilot A, Taghzouti K, Simon H, Le Moal M. Limbic system, basal ganglia, and dopaminergic neurons. Executive and regulatory neurons and their role in the organization of behavior. Brain Behav Evol 1989; 33(2-3): 157-61.
        [http://dx.doi.org/10.1159/000115920] [PMID: 2758295]

[2]   Schultz W, Apicella P, Scarnati E, Ljungberg T. Neuronal activity in monkey ventral striatum related to the expectation of reward. J Neurosci 1992; 12(12): 4595-610.
        [http://dx.doi.org/10.1523/JNEUROSCI.12-12-04595.1992] [PMID: 1464759]

[3]   Apicella P, Ljungberg T, Scarnati E, Schultz W. Responses to reward in monkey dorsal and ventral striatum. Exp Brain Res 1991; 85(3): 491-500.
        [http://dx.doi.org/10.1007/BF00231732] [PMID: 1915708]

[4]   Wang JL, Wang B, Chen W. Differences in cocaine-induced place preference persistence, locomotion and social behaviors between C57BL/6J and BALB/cJ mice. Dongwuxue Yanjiu 2014 Sep; 35(5): 426-35.
        [http://dx.doi.org/10.13918/j.issn.2095-8137.2014.5.426] [PMID: 25297083] [PMCID: PMC4790360]

[5]   Kalivas PW. Cocaine and amphetamine-like psychostimulants: Neurocircuitry and glutamate neuroplasticity. Dialogues Clin Neurosci 2007; 9(4): 389-97.
        [http://dx.doi.org/10.31887/DCNS.2007.9.4/pkalivas] [PMID: 18286799]

[6]   Niu F, Liao K, Hu G, et al. Cocaine-induced release of CXCL10 from pericytes regulates monocyte transmigration into the CNS. J Cell Biol 2019; 218(2): 700-21.
        [http://dx.doi.org/10.1083/jcb.201712011] [PMID: 30626719]

[7]   Sordo L, Indave BI, Barrio G, Degenhardt L, de la Fuente L, Bravo MJ. Cocaine use and risk of stroke: A systematic review. Drug Alcohol Depend 2014; 142: 1-13.
        [http://dx.doi.org/10.1016/j.drugalcdep.2014.06.041] [PMID: 25066468]

[8]   Planeta CS, Lepsch LB, Alves R, Scavone C. Influence of the dopaminergic system, CREB, and transcription factor-κB on cocaine neurotoxicity. Braz J Med Biol Res 2013; 46(11): 909-15.
        [http://dx.doi.org/10.1590/1414-431X20133379] [PMID: 24141554]

[9]   Dietrich JB, Mangeol A, Revel MO, Burgun C, Aunis D, Zwiller J. Acute or repeated cocaine administration generates reactive oxygen species and induces antioxidant enzyme activity in dopaminergic rat brain structures. Neuropharmacology 2005; 48(7): 965-74.
        [http://dx.doi.org/10.1016/j.neuropharm.2005.01.018] [PMID: 15857623]

[10]   Hammer RP Jr, Cooke ES. Gradual tolerance of metabolic activity is produced in mesolimbic regions by chronic cocaine treatment, while subsequent cocaine challenge activates extrapyramidal regions of rat brain. J Neurosci 1994; 14(7): 4289-98.
        [http://dx.doi.org/10.1523/JNEUROSCI.14-07-04289.1994] [PMID: 8027779]

[11]   Zocchi A, Conti G, Orzi F. Differential effects of cocaine on local cerebral glucose utilization in the mouse and in the rat. Neurosci Lett 2001; 306(3): 177-80.
        [http://dx.doi.org/10.1016/S0304-3940(01)01898-5] [PMID: 11406324]

[12] Guha Prasun. Cocaine elicits autophagic cytotoxicity *via* a nitric oxide-GAPDH signaling cascade. Biol Sci 2016; 113(5): 1417-22.

[13] Wise RA. Dopamine, learning and motivation. Nat Rev Neurosci 2004; 5(6): 483-94.
[http://dx.doi.org/10.1038/nrn1406] [PMID: 15152198]

[14] Kolb B, Pellis S, Robinson TE. Plasticity and functions of the orbital frontal cortex. Brain Cogn 2004; 55(1): 104-15.
[http://dx.doi.org/10.1016/S0278-2626(03)00278-1] [PMID: 15134846]

[15] Ramirez ID. Cocaine-induced synaptic redistribution of nmdars in striatal neurons alters NMDAR-dependent signal transduction. Front Neurosci 2020; 698.

[16] Aguinaga D, Medrano M. Cocaine effects on dopaminergic transmission depend on a balance between sigma-1 and sigma-2 receptor expression. Front Mol Neurosci 2018; 12.

[17] Zhao ZD, Han X. A molecularly defined D1 medium spiny neuron subtype negatively regulates cocaine addiction. Sci Adv 2022; 8 (32).

[18] Ungless MA, Whistler JL, Malenka RC, Bonci A. Single cocaine exposure *in vivo* induces long-term potentiation in dopamine neurons. Nature 2001; 411(6837): 583-7.
[http://dx.doi.org/10.1038/35079077] [PMID: 11385572]

[19] Bellone C, Lüscher C. Cocaine triggered AMPA receptor redistribution is reversed *in vivo* by mGluR-dependent long-term depression. Nat Neurosci 2006; 9(5): 636-41.
[http://dx.doi.org/10.1038/nn1682] [PMID: 16582902]

[20] Yuan T, Mameli M, O'Connor EC, *et al.* Expression of cocaine-evoked synaptic plasticity by GluN3A-containing NMDA receptors. Neuron 2013; 80(4): 1025-38.
[http://dx.doi.org/10.1016/j.neuron.2013.07.050] [PMID: 24183704]

[21] Chen BT, Bowers MS, Martin M, *et al.* Cocaine but not natural reward self-administration nor passive cocaine infusion produces persistent LTP in the VTA. Neuron 2008; 59(2): 288-97.
[http://dx.doi.org/10.1016/j.neuron.2008.05.024] [PMID: 18667156]

[22] Creed M, Kaufling J, Fois GR, *et al.* Cocaine exposure enhances the activity of ventral tegmental area dopamine neurons *via* calcium-impermeable NMDARs. J Neurosci 2016; 36(42): 10759-68.
[http://dx.doi.org/10.1523/JNEUROSCI.1703-16.2016] [PMID: 27798131]

[23] Dong Y, Saal D, Thomas M, *et al.* Cocaine-induced potentiation of synaptic strength in dopamine neurons: Behavioral correlates in GluRA(–/–) mice. Proc Natl Acad Sci 2004; 101(39): 14282-7.
[http://dx.doi.org/10.1073/pnas.0401553101] [PMID: 15375209]

[24] Vanderschuren LJMJ, Kalivas PW. Alterations in dopaminergic and glutamatergic transmission in the induction and expression of behavioral sensitization: A critical review of preclinical studies. Psychopharmacology 2000; 151(2-3): 99-120.
[http://dx.doi.org/10.1007/s002130000493] [PMID: 10972458]

[25] Hamilton CJ, Lim RL. Chromatin-mediated alternative splicing regulates cocaine-reward behaviour. Neuron 2021; 109(18): 2943-66.

[26] Schmidt HD, Pierce RC. Cocaine-induced neuroadaptations in glutamate transmission. Ann N Y Acad Sci 2010; 1187(1): 35-75.
[http://dx.doi.org/10.1111/j.1749-6632.2009.05144.x] [PMID: 20201846]

[27] Mantsch JR, Vranjkovic O, Twining RC, Gasser PJ, McReynolds JR, Blacktop JM. Neurobiological mechanisms that contribute to stress-related cocaine use. Neuropharmacology 2014; 76(Part B): 383-94.
[http://dx.doi.org/10.1016/j.neuropharm.2013.07.021]

[28] Lucantonio F, Stalnaker TA, Shaham Y, Niv Y, Schoenbaum G. The impact of orbitofrontal dysfunction on cocaine addiction. Nat Neurosci 2012; 15(3): 358-66.
[http://dx.doi.org/10.1038/nn.3014] [PMID: 22267164]

[29]   Kennedy PJ, Feng J, Robison AJ, *et al.* Class I HDAC inhibition blocks cocaine-induced plasticity by targeted changes in histone methylation. Nat Neurosci 2013; 16(4): 434-40.
[http://dx.doi.org/10.1038/nn.3354] [PMID: 23475113]

[30]   Anier K, Malinovskaja K, Aonurm-Helm A, Zharkovsky A, Kalda A. DNA methylation regulates cocaine-induced behavioral sensitization in mice. Neuropsychopharmacology 2010; 35(12): 2450-61.

[31]   LaPlant Q, Vialou V, Covington HE, *et al.* Dnmt3a regulates emotional behavior and spine plasticity in the nucleus accumbens. Nat Neurosci 2010; 13(9): 1137-43.
[http://dx.doi.org/10.1038/nn.2619] [PMID: 20729844]

[32]   Kumar A, Choi KH, Renthal W, *et al.* Chromatin remodeling is a key mechanism underlying cocaine-induced plasticity in striatum. Neuron 2005; 48(2): 303-14.
[http://dx.doi.org/10.1016/j.neuron.2005.09.023] [PMID: 16242410]

[33]   Levine AA, Guan Z, Barco A, Xu S, Kandel ER, Schwartz JH. CREB-binding protein controls response to cocaine by acetylating histones at the fosB promoter in the mouse striatum. Proc Natl Acad Sci 2005; 102(52): 19186-91.
[http://dx.doi.org/10.1073/pnas.0509735102] [PMID: 16380431]

[34]   Brami-Cherrier K, Valjent E, Hervé D, *et al.* Parsing molecular and behavioral effects of cocaine in mitogen- and stress-activated protein kinase-1-deficient mice. J Neurosci 2005; 25(49): 11444-54.
[http://dx.doi.org/10.1523/JNEUROSCI.1711-05.2005] [PMID: 16339038]

[35]   Matsui A, Alvarez VA. Cocaine inhibition of synaptic transmission in the ventral pallidum is pathway-specific and mediated by serotonin. Cell Rep 2018; 23(13): 3852-63.
[http://dx.doi.org/10.1016/j.celrep.2018.05.076] [PMID: 29949769]

[36]   Francis TC, Gantz SC, Moussawi K, Bonci A. Synaptic and intrinsic plasticity in the ventral tegmental area after chronic cocaine. Curr Opin Neurobiol 2019 Feb; 54: 66-72. Epub 2018 Sep 17.
[http://dx.doi.org/10.1016/j.conb.2018.08.013] [PMID: 30237117] [PMCID: PMC10131346]

[37]   Kauer JA, Malenka RC. Synaptic plasticity and addiction. Nat Rev Neurosci 2007; 8(11): 844-58.
[http://dx.doi.org/10.1038/nrn2234] [PMID: 17948030]

[38]   Dong Y, Nestler EJ. The neural rejuvenation hypothesis of cocaine addiction. Trends Pharmacol Sci 2014; 35(8): 374-83.
[http://dx.doi.org/10.1016/j.tips.2014.05.005] [PMID: 24958329]

[39]   Block ER, Nuttle J, Balcita-Pedicino JJ, *et al.* Brain region-specific trafficking of the dopamine transporter. J Neurosci 2015; 35(37): 12845-58.
[http://dx.doi.org/10.1523/JNEUROSCI.1391-15.2015] [PMID: 26377471]

[40]   Lange RA, Hillis LD. Cardiovascular complications of cocaine use. N Engl J Med 2001; 345(5): 351-8.
[http://dx.doi.org/10.1056/NEJM200108023450507] [PMID: 11484693]

[41]   Zhang XF, Hu XT, White FJ. Whole-cell plasticity in cocaine withdrawal: Reduced sodium currents in nucleus accumbens neurons. J Neurosci 1998; 18(1): 488-98.
[http://dx.doi.org/10.1523/JNEUROSCI.18-01-00488.1998] [PMID: 9412525]

[42]   Perez MF, Ford KA, Goussakov I, Stutzmann GE, Hu XT. Repeated cocaine exposure decreases dopamine D_2 -like receptor modulation of Ca^{2+} homeostasis in rat nucleus accumbens neurons. Synapse 2011; 65(2): 168-80.
[http://dx.doi.org/10.1002/syn.20831] [PMID: 20665696]

[43]   Hu XT, Ford K, White FJ. Repeated cocaine administration decreases calcineurin (PP2B) but enhances DARPP-32 modulation of sodium currents in rat nucleus accumbens neurons. Neuropsychopharmacology 2005; 30(5): 916-26.
[http://dx.doi.org/10.1038/sj.npp.1300654] [PMID: 15726118]

[44]   Zhang XF, Cooper DC, White FJ. Repeated cocaine treatment decreases whole-cell calcium current in

rat nucleus accumbens neurons. J Pharmacol Exp Ther 2002; 301(3): 1119-25.
[http://dx.doi.org/10.1124/jpet.301.3.1119] [PMID: 12023545]

[45]   Wei C, Han X, Weng D, *et al.* Response dynamics of midbrain dopamine neurons and serotonin neurons to heroin, nicotine, cocaine, and MDMA. Cell Discov 2018; 4(1): 60.
[http://dx.doi.org/10.1038/s41421-018-0060-z] [PMID: 30416749]

[46]   Fattore L, Puddu MC, Picciau S, *et al.* Astroglial *in vivo* response to cocaine in mouse dentate gyrus: a quantitative and qualitative analysis by confocal microscopy. Neuroscience 2002; 110(1): 1-6.
[http://dx.doi.org/10.1016/S0306-4522(01)00598-X] [PMID: 11882367]

[47]   Li Z, Gao L, Liu Q, *et al.* Aquaporin-4 knockout regulated cocaine-induced behavior and neurochemical changes in mice. Neurosci Lett 2006; 403(3): 294-8.
[http://dx.doi.org/10.1016/j.neulet.2006.05.004] [PMID: 16797122]

[48]   Messer CJ, Eisch AJ, Carlezon WA Jr, *et al.* Role for GDNF in biochemical and behavioral adaptations to drugs of abuse. Neuron 2000; 26(1): 247-57.
[http://dx.doi.org/10.1016/S0896-6273(00)81154-X] [PMID: 10798408]

[49]   Kumar S, Crenshaw BJ, Williams SD, Bell CR, Matthews QL, Sims B. Cocaine-specific effects on exosome biogenesis in microglial cells. Neurochem Res 2021; 46(4): 1006-18.
[http://dx.doi.org/10.1007/s11064-021-03231-2] [PMID: 33559104]

[50]   Garg UC, Turndorf H, Bansinath M. Effect of cocaine on macromolecular syntheses and cell proliferation in cultured glial cells. Neuroscience 1993; 57(2): 467-72.
[http://dx.doi.org/10.1016/0306-4522(93)90079-U] [PMID: 7509470]

[51]   Sharpe AL, Trzeciak M, Eliason NL, *et al.* Repeated cocaine or methamphetamine treatment alters astrocytic CRF2 and GLAST expression in the ventral midbrain. Addict Biol 2022; 27(2): e13120.
[http://dx.doi.org/10.1111/adb.13120] [PMID: 34825430]

[52]   Berhow MT, Hiroi N, Kobierski LA, Hyman SE, Nestler EJ. Influence of cocaine on the JAK-STAT pathway in the mesolimbic dopamine system. J Neurosci 1996; 16(24): 8019-26.
[http://dx.doi.org/10.1523/JNEUROSCI.16-24-08019.1996] [PMID: 8987828]

[53]   Berhow MT, Hiroi N, Nestler EJ. Regulation of ERK (extracellular signal regulated kinase), part of the neurotrophin signal transduction cascade, in the rat mesolimbic dopamine system by chronic exposure to morphine or cocaine. J Neurosci 1996; 16(15): 4707-15.
[http://dx.doi.org/10.1523/JNEUROSCI.16-15-04707.1996] [PMID: 8764658]

[54]   Beitner-Johnson D, Guitart X, Nestler EJ. Neurofilament proteins and the mesolimbic dopamine system: common regulation by chronic morphine and chronic cocaine in the rat ventral tegmental area. J Neurosci 1992; 12(6): 2165-76.
[http://dx.doi.org/10.1523/JNEUROSCI.12-06-02165.1992] [PMID: 1376774]

[55]   Beitner-Johnson D, Nestler EJ. Morphine and cocaine exert common chronic actions on tyrosine hydroxylase in dopaminergic brain reward regions. J Neurochem 1991; 57(1): 344-7.
[http://dx.doi.org/10.1111/j.1471-4159.1991.tb02133.x] [PMID: 1675665]

[56]   Nicolucci C, Pais ML, Santos AC, *et al.* Single low dose of cocaine–structural brain injury without metabolic and behavioral changes. Front Neurosci 2021; 14: 589897.
[http://dx.doi.org/10.3389/fnins.2020.589897] [PMID: 33584173]

[57]   Kohut SJ, Mintzopoulos D, Kangas BD, *et al.* Effects of long-term cocaine self-administration on brain resting-state functional connectivity in nonhuman primates. Transl Psychiatry 2020; 10(1): 420.
[http://dx.doi.org/10.1038/s41398-020-01101-z] [PMID: 33268770]

CHAPTER 2

Heroin and its Effects on the Brain

Abstract: As per the Drug Enforcement Administration (DEA) and the Controlled Substances Act (CSA) reports, heroin is a Schedule I drug. Heroin causes addiction to the brain like any other addictive substance. Heroin addiction has both long-term and short-term effects on the body. The brain has natural opioid receptors. Heroin is a synthetic opioid. When taken regularly, the brain stops making its own natural opioids. This affects the pain/reward system and causes withdrawal symptoms in patients. Heroin addiction damages the brain's reward system and breathing. Less breathing causes less oxygen supply to the brain. There are reports that state that dementia-like situation is created in the brain due to heroin abuse. Heroin lipophilicity allows the entry of it into the Blood Brain Barrier. μ-opioid receptors (MOR), causing the addictive effects of the heroin in the brain. Dementia symptoms, memory issues, and mental health changes like depression or anxiety are the symptoms that are caused by heroin abuse. Both individual and environmental factors influence a person's ability to abuse heroinanopioid which provides intense feelings of pleasure.

Keywords: Anxiety, Dementia symptoms, Depression, Heroin, Memory issues, Opioid, μ-opioid receptor (MOR).

INTRODUCTION

Heroin use affects not only neurotransmitters but also the hormonal systems in an irreversible way [1, 2]. Human studies have shown white matter deterioration in heroin users [3, 4]. Physical dependence and tolerance are the two effects of heroin use. Withdrawal of the drug is caused within hours of the last drug ingestion characterised by muscle and bone pain, vomiting, insomnia, diarrhoea, restlessness, and cold flashes. Heroin use disorder is caused by repeated heroin use and it is accompanied by uncontrolled drug seeking. Needle sharing and injection is the main cause of HIV in patients with heroin use. The other dangerous effect of using Heroin in pregnant mothers is that it crosses the placenta and it makes the baby in the womb to be dependent on the drug called neonatal abstinence syndrome (NAS). Overdose of heroin can be dangerous to the life-threatening cause and requires medical attention. Almost 75% of heroin users are identified with mental health issues such as Borderline personality disorder, ADHD, depression, or bipolar disorder [5 - 7]. Many people with substance use disorder may have mental health issues or *vice versa*. Research also points out that co-occurring mental illnesses are common in adolescents with substance use disorders. Generalized anxiety disorder, panic disorder, and posttraumatic stress disorder are widespread in co-morbid substance use disorder. Heroin is classified as Schedule I drug as per the regulations of the Drug Enforcement Administration and Controlled Substances Act. After taking heroin immediately, the users feel high pleasure. Between 4 to 5 hrs, this effect is seen. Based on the methods of administration, this effect can be seen. For example, it can be seen within 20 seconds, peaking after 2 hours and up to 4 hours. Heroin is a very addictive opioid that decreases pain, induces euphoria and warmth, and causes drowsiness. Naloxone is the medicine which is used to treat heroin overdose. From 2007, the use of heroins among people is increasing. Medication-assisted treatment (MAT) is the treatment for heroin addiction. Methadone and buprenorphine can help withdrawal symptoms. Smoking or snorting can cause the high within about 10 minutes and can persist 4 to 5 hours.

Heroin and the Brain

Pain perception mechanisms are modulated by heroin use. Heroin belongs to illicit opioid drugs that can become morphine-like substances when entered into the brain. Heroin, upon entry into the brain, binds to opioid receptors in the brain stem, and the body. This heroin opioid inhibits gamma-aminobutyric acid (GABA), upon doing it, the level of dopamine is increased in brain. The reinforcing property of heroin is due to the secretion of Dopamine at excess levels. Responding to stressful stimuli, managing behaviour, and decision-making is affected by heroin use due to white matter decay [8]. The basal ganglia, prefrontal cortex, extended amygdala are the regions affected by drug abuse. Reward circuit in the basal ganglia is over activated by drugs causing euphoria of the drug high. The basal ganglia play an important role in positive forms of motivation. Repeated exposure to a drug causes this pathway to be adapted to the drug, reducing the sensitivity to the need for more doses of the drug to get the same feeling. Anxiety, irritability, and unease are stress feelings which are caused by withdrawal of the drug Opioid can cause death if taken overdose as it causes decreased breathing. These effects are due to the changes in brain stem which controls heart rate, breathing, and sleeping. Overdose of heroin causes hypoxia that leads to short-term and long-term effects on neuronal cells. Repeated heroin use causes heroin use disorder leading to uncontrolled drug seeking.

Heroin is extremely addictive; withdrawal begins within 5 hours.

Action of Heroin on Mu-opioid Receptors

Mu-opioid receptors (MoRs) are considered the reward system of the brain. When dopamine binds to these receptors, it creates a sense of well-being or pleasure.

Heroin binds to these receptors, thus mimicking the effect of dopamine. Heroin also damages the white matter, thus, in turn, affecting people's decision-making ability, self-control, and ability to face difficult situations in life. Structural changes are also noticed in the brain due to heroin usage such as depression, paranoia, and psychosis. Venus sclerosis is also one of the effects of the usage of heroins, thus affecting venous circulation. An unnatural sense of well-being is one of the short-term effect of heroin usage.

Mental confusion and difficulty in thinking were also noticed as short-term changes in heroin usage. Unnatural mood swings, depression, insomnia, paranoia, psychosis are the long-term effects of heroin usage. Upon entry into the brain, heroin is converted as morphine. Not only heroins but prescription drugs such as OxyContin (oxycodone), Vicodin (acetaminophen/hydrocodone), fentanyl, methadone, and Dilaudid (hydromorphone) can gel with opioid receptors and cause the release of dopamine.

There was a study on the brains of postpartum samples of victims who died of heroin abuse. Their brains have shown tremendous changes in the neurogensis process. Such changes are noted by changes in reduced neural progenitor cell number, cell proliferation rates were low, and there was also less number of dendritic trees [9]. Cynomolgus macaques have shown a different gene expression pattern in response to heroin administration in the hippocampus and striatum at different points of their age. Upon treatment with heroin, the animals have shown differential expression in genes related to dopamine, synapse, autophagy, and neurotrophin signalling [10, 11].

CONCLUSION

Heroin use disorder is caused by repeated heroin use and it is accompanied by uncontrolled drug seeking. Needle sharing and injection is the main cause of HIV in patients with heroin use. The other dangerous effect of using Heroin in pregnant mothers is that it crosses the placenta and makes the baby in the womb dependent on the drug called neonatal abstinence syndrome (NAS). Overdose of heroin can be dangerous to the life-threatening cause and requires medical attention. Pain perception mechanisms are modulated by heroin use. Heroin belongs to illicit opioid drugs that can become morphine-like substances when entered into the brain. Heroin, upon entry into the brain, binds to opioid receptors in the brain stem, and the body. This heroin opioid inhibits gamma-aminobutyric acid (GABA); upon doing it, the level of dopamine is increased in the brain. Mental confusion and difficulty in thinking were also noticed as short-term changes in heroin usage. Unnatural mood swings, depression, insomnia, paranoia, psychosis are the long-term effects of heroin usage. Upon entry into the brain, heroin is converted to morphine.

REFERENCES

[1] Ignar DM, Kuhn CM. Effects of specific mu and kappa opiate tolerance and abstinence on hypothalamo-pituitary-adrenal axis secretion in the rat. J Pharmacol Exp Ther 1990; 255(3): 1287-95.
[PMID: 2175800]

[2] Kreek MJ, Ragunath J, Plevy S, Hamer D, Schneider B, Hartman N. ACTH, cortisol and β-endorphin response to metyrapone testing during chronic methadone maintenance treatment in humans. Neuropeptides 1984; 5(1-3): 277-8.
[http://dx.doi.org/10.1016/0143-4179(84)90081-7] [PMID: 6099512]

[3] Li W, Li Q, Zhu J, *et al.* White matter impairment in chronic heroin dependence: A quantitative DTI study. Brain Res 2013; 1531: 58-64.
[http://dx.doi.org/10.1016/j.brainres.2013.07.036] [PMID: 23895765]

[4] Qiu Y, Jiang G, Su H, *et al.* Progressive white matter microstructure damage in male chronic heroin dependent individuals: A DTI and TBSS study. PLoS One 2013; 8(5): e63212.
[http://dx.doi.org/10.1371/journal.pone.0063212] [PMID: 23650554]

[5] Kreek MJ, Levran O, Reed B, Schlussman SD, Zhou Y, Butelman ER. Opiate addiction and cocaine addiction: Underlying molecular neurobiology and genetics. J Clin Invest 2012; 122(10): 3387-93.
[http://dx.doi.org/10.1172/JCI60390] [PMID: 23023708]

[6] Conway KP, Compton W, Stinson FS, Grant BF. Lifetime comorbidity of DSM-IV mood and anxiety disorders and specific drug use disorders: Results from the national epidemiologic survey on alcohol and related conditions. J Clin Psychiatry 2006; 67(2): 247-58.
[http://dx.doi.org/10.4088/JCP.v67n0211] [PMID: 16566620]

[7] Torrens M, Gilchrist G, Domingo-Salvany A. Psychiatric comorbidity in illicit drug users: Substance-induced *versus* independent disorders. Drug Alcohol Depend 2011; 113(2-3): 147-56.
[http://dx.doi.org/10.1016/j.drugalcdep.2010.07.013] [PMID: 20801586]

[8] Compton WM, Thomas YF, Stinson FS, Grant BF. Prevalence, correlates, disability, and comorbidity of DSM-IV drug abuse and dependence in the United States: Results from the national epidemiologic survey on alcohol and related conditions. Arch Gen Psychiatry 2007; 64(5): 566-76.
[http://dx.doi.org/10.1001/archpsyc.64.5.566] [PMID: 17485608]

[9] Sun Y, Wang GB, Lin QX. Disrupted white matter structural connectivity in heroin abusers. Addict Biol 2015; 22(1): 184-95.
[PMID: 26177615]

[10] Bayer R, Franke H, Ficker C, *et al.* Alterations of neuronal precursor cells in stages of human adult neurogenesis in heroin addicts. Drug Alcohol Depend 2015; 156: 139-49.
[http://dx.doi.org/10.1016/j.drugalcdep.2015.09.005] [PMID: 26416695]

[11] Choi MR, Jin YB, Bang SH, *et al.* Age-related effects of heroin on gene expression in the hippocampus and striatum of cynomolgus monkeys. Clin Psychopharmacol Neurosci 2020; 18(1): 93-108.
[http://dx.doi.org/10.9758/cpn.2020.18.1.93] [PMID: 31958910]

CHAPTER 3

MDMA (3,4-methylenedioxy-methamphetamine)

Abstract: 3,4-methylenedioxy-methamphetamine (MDMA) is a synthetic drug very similar to hallucinogens and stimulants. This drug is also called ecstasy or molly. It produces feelings of pleasure, warmth, distorted sensory time and perception. MDMA increases the activity of serotonin, dopamine and norepinephrine in the brain. It causes various health effects such as nausea, sweating, chills, muscle cramping, *etc.* The effect of this drug can be seen in 3 to 5 hours in the body. A spike in body temperature can be seen in MDMA users that can be fatal as it affects the liver, kidney, and heart leading to death. Addiction to MDMA is not yet proven, however, withdrawal symptoms such as fatigue and depression are noted. MDMA is usually taken *via* the mouth or snorting in the form of tablets or capsules. This drug is also taken or abused along with other drugs such as LSD, alcohol, and marijuana. MDMA is a scheduled drug with no proven medical use. MDMA causes a surge of serotonin, dopamine, and norepinephrine in the brain to regulate mood, learning, memory, stress, anxiety, *etc.* This chapter discusses the effects of MDMA on the human brain.

Keywords: Dopamine, Memory, MDMA on the human brain, Norepinephrine learning.

INTRODUCTION

In healthy young people, the use of MDMA can lead to cognitive decline when abused with cannabis. Impairment of working memory is considered a cognitive decline parameter [1]. Plasma membrane serotonin transporter (SERT), is reduced in heavy MDMA users [2]. Dose-dependent reductions in 5-HT were seen in all brain regions of rats upon MDMA administration. Norepinephrine and/or dopamine levels are also reduced in some of the brain regions. In the guinea pig brain 20mg/kg dose of MDMA reduced 5-HT in all brain regions [3]. In MDMA-treated monkeys, even after 7 years, abnormal brain 5-HT innervation patterns were seen [4]. Losses of serotonergic (5-HT) axons are seen throughout the forebrain during 20 mg/kg, s.c., twice daily for 4 d in rat brain. Logohhera psychomotor drive and enhanced insight are noted in subjects along with toxic psychosis [5]. METH and cocaine are classified as stimulants, however, there are key differences between them. The key differences include drug type, classification and origination, process of metabolization/half-life, appearance, and effects on the brain and body. METH is a man-made synthetic chemical; however, cocaine is synthesised from the coca plant. These two drugs also differ in their half-life, for cocaine, the half-life is one hour but for METH half-life is 12 hours. Coke appears as a fine white powder, whereas METH is appear like crystal glass powder. Both drugs block the reuptake of dopamine in the brain, however, METH also increases the release of dopamine in the brain contributing to more dopamine levels. The addictive potential of METH is less understood than other drugs.

Norepinephrine (NE) and MDMA

Norepinephrine (NE) transporter inhibitor reboxetine has been observed to reduce the BP heart rate and high feeling. NE levels, stimulation, and excitation in healthy subjects [6]. MDMA causes hyponatremia and hyponatremia-associated deaths [7]. Tryptophan hydroxylase is released due to oxidative stress due to MDMA, leaving the cell vulnerable to oxidative stress. MDMA also releases norepinephrine, dopamine, and acetylcholine release and acts on various receptors such as alpha 2-adrenergic and 5-hydroxytryptamine (5-HT) 2A receptors. Receptors such as 5-HT2 and 5-HT1 are known to play a role in MDMA responses. Among these receptors, 5-HT2 receptors mediate positive mood but not negative mood [8]. MDMA is proven to act through serotonin receptor binding with high affinity [9].

In squirrel monkeys, which is a human primates, it is shown that a single oral dose of MDMA (5.7 mg/kg) has created long-lasting serotonergic defects [10]. Radioligand carbon-11-labeled McN-5652 in the human brain under PET study has shown that 5 HT binding was decreased [10]. In MDMA users, 5-HT transporter (SERT) reduced binding in multiple brain regions and DAT binding [11]. Cerebral presynaptic serotonergic transmitter system was changed in human young adults as revealed by a positron emission tomography study [12].

Ketanserin,5-HT2 receptor antagonist, has been shown to decrease the effect of MDMA on emotional excitation and perception [13]. Doxazosin, an α_1-noradrenergic receptor antagonist, was shown to be cardiogeneic and thermogenic, contributing to the euphoric effects of MDMA in humans [14]. The stimulant effect of MDMA is reduced by reboxetine, a norepinephrine transporter inhibitor. Moreover, MDMA plays a role in increasing or decreasing the body temperature and MDMA is an alpha-adrenoceptor agonist like clonidine which produces hypothermia [15]. Plasma oxytocin levels were increased to a peak of 83.7 pg/ml after MDMA (1.5 mg/kg) at 90-120min time in healthy volunteers with MDMA use [16]. Emotional empathy was noticed in MDMA doses without affecting cognitive empathy [17].

Antioxidant and Signalling Pathways and MDMA

The brain's antioxidant system is also affected due to MDMA treatment. For example, in male rats after MDMA treatment, superoxide dismutase levels (SOD1) were increased in the female brain whereas SOD2 was increased in the male brain. It is also found that the male brain is very sensitive to the effects of MDMA-induced neurotoxic effects [18]. In rat cortical neurons treated with MDMA, it was found that under the influence of MDMA, the autophagy process is ii dated with the activation of AMPK/ULK1 signaling pathway [19]. MDMA also triggers autophagy in serotonergic neurons in cultured rat neurons [20 - 22]. Behavioural sensitization in C57BL/6J mice is due to the changes in the properties of long-term potentiation in serotonergic and noradrenergic neurons after repeated MDMA exposure [23]. Rat cortical embryonic cortical primary cell cultures were exposed to MDMA and it was found that stem cells and neurons were decreased [24]. Cell viability and cytoskeletal light filament were reduced in rat raphe nuclear cell lines upon treatment with MDMA, but BDNF seems to protect these neurons from the damaging effect [25].

Non-human, Primates, and MDMA Effects

Experiments conducted with MDMA on human primates, *i.e.*, on monkeys for three weeks have reduced the level of presynaptic serotonin markers [26]. Differential electrophysiological firings were noted in brain slices of ventral midbrain dopaminergic neurons exposed to different concentrations of MDMA. At lower concentrations (1 mol/L) MDMA excited the cells, whereas at higher concentrations (10 –30 mol/L) it caused the hyperpolarization of the cell. These higher concentration effects are due to the effects caused by the activation of D2 autoreceptors [27]. The same observation is seen in rats exposed to different concentrations of MDMA. In the hippocampus, there was a significant decrease of neurons in MDMA-exposed rats in comparison to the control and this is due to different dose treatments of MDMA [28]. People may experience trouble concentrating, impaired memory, judgement, difficulty, unable to recognise dangerous situations, and unable to process information in the brain due to the long time effects of heroin. Learning and memory were affected in developing animals which are exposed to MDMA. In mice, MDMA has been shown to cause the loss of dopaminergic cells in the substantia nigra [29].

Mitochondria and MDMA

Mitochondrial trafficking and mitochondrial fragmentation were increased in the hippocampus due to MDMA treatment, indicating mitochondrial involvement [30]. In contrast, the recreational use of MDMA is not doing long-term changes on serotonergic neurons [31]. Compared to control the MDMA-treated group had a significant reduction in the distribution volume ratio in the mesencephalon [32]. However, in neonatal rat brains, a single dose of MDMA has caused neurodegeneration [33]. MDMA increases serotonin, dopamine, and noradrenaline levels in the brain [34].

MDMA and Hippocampus

In male Wistar rats, 10mg/kg of MDMA increased apoptosis and caused different anxiety-like behaviors [35]. In monkeys, even after 7 years of MAMA treatment, there were altered serotonin levels [36]. Addictive behaviour in mice is mediated by long-term administration of MDMA, causing strong neural adaptations. MDMA-induced neurotoxicity is also due to iNOS-derived nitrosative stress in rats [37].WAY100635 is a 5HT receptor antagonist, pretreatment with this in rats causes the reduced cFOS expression upon MDMA treatment in oxycontin neurons. MDMA increases cannabeniod receptors in the rat brain, which also causes serotonergic neurotoxicity in rats [38]. MDMA treatment in the hippocampus led to NFKB activation, IL1 beta release and microglial activation [39]. In the hypothalamus and cortex, MDMA has increased the levels of IL1beta and serotonin depletion in rats [40].

Inflammatory Cytokines and MDMA

Further studies on this line have indicated that received MDMA (12.5 mg/kg, i.p.), has shown increased IL-1ra levels and decreased IL1-RI expression in cell bodies of neurons [41]. In chronic MDMA users with vascular heart disease in widespread which is due to the metabolite of MDMA known as 3,4-methylenedioxyamphetamine (MDA) [42]. MDMA users will have reduced cerebrospinal fluid 5-hydroxyindoleacetic acid and brain 5-HT transporters leading to cognitive dysfunctions (McCann *et al.*, 2000). Using positron emission tomography in humans, it is shown that MDMA induces 5HT neurotoxicity [43, 44].

CONCLUSION

In healthy young people, the use of MDMA can lead to cognitive decline when abused with cannabis. Impairment of working memory is considered a cognitive decline parameter. The brain's antioxidant system is also affected due to MDMA treatment. For example, in male rats after MDMA treatment, superoxide dismutase levels (SOD1) were increased in the female brain whereas SOD2 was increased in the male brain. It is also found that the male brain is very sensitive to the effects of MDMA-induced neurotoxic effects. Experiments conducted with MDMA on human primates, *i.e.*, on monkeys for three weeks have reduced the level of presynaptic serotonin markers. Differential electrophysiological firings were noted in brain slices of ventral midbrain dopaminergic neurons exposed to different concentrations of MDMA. Mitochondrial trafficking and mitochondrial fragmentation were increased in the hippocampus due to MDMA treatment, indicating mitochondrial involvement.

REFERENCES

[1]   Gouzoulis-Mayfrank E, Daumann J, Tuchtenhagen F, *et al.* Impaired cognitive performance in drug free users of recreational ecstasy (MDMA). J Neurol Neurosurg Psychiatry 2000; 68(6): 719-25.
       [http://dx.doi.org/10.1136/jnnp.68.6.719] [PMID: 10811694]

[2]   Biezonski DK, Meyer JS. The nature of 3, 4-methylenedioxymethamphetamine (MDMA) induced serotonergic dysfunction: Evidence for and against the neurodegeneration hypothesis. Curr Neuropharmacol 2011; 9(1): 84-90.
       [http://dx.doi.org/10.2174/157015911795017146] [PMID: 21886568]

[3]   Commins DL, Vosmer G, Virus RM, Woolverton WL, Schuster CR, Seiden LS. Biochemical and histological evidence that methylenedioxymethylamphetamine (MDMA) is toxic to neurons in the rat brain. J Pharmacol Exp Ther 1987; 241(1): 338-45.
       [PMID: 2883295]

[4]   Hatzidimitriou G, McCann UD, Ricaurte GA. Altered serotonin innervation patterns in the forebrain of monkeys treated with (+/-)3,4-methylenedioxymethamphetamine seven years previously: Factors influencing abnormal recovery. J Neurosci 1999; 19(12): 5096-107.
       [http://dx.doi.org/10.1523/JNEUROSCI.19-12-05096.1999] [PMID: 10366642]

[5]   Hermle L, Spitzer M, Borchardt D, Kovar KA, Gouzoulis E. Psychological effects of MDE in normal subjects. Are entactogens a new class of psychoactive agents? Neuropsychopharmacology 1993; 8(2): 171-6.
       [http://dx.doi.org/10.1038/npp.1993.19] [PMID: 8471129]

[6]   Hysek CM, Simmler LD, Ineichen M, *et al.* The norepinephrine transporter inhibitor reboxetine reduces stimulant effects of MDMA ("ecstasy") in humans. Clin Pharmacol Ther 2011; 90(2): 246-55.
       [http://dx.doi.org/10.1038/clpt.2011.78] [PMID: 21677639]

[7]   Campbell GA, Rosner MH. The agony of ecstasy: MDMA (3,4-methylenedioxymethamphetamine) and the kidney. Clin J Am Soc Nephrol 2008; 3(6): 1852-60.
       [http://dx.doi.org/10.2215/CJN.02080508] [PMID: 18684895]

[8]   Van Wel JHP, Kuypers KPC, Theunissen EL, Bosker WM, Bakker K, Ramaekers JG. Effects of acute MDMA intoxication on mood and impulsivity: role of the 5-HT2 and 5-HT1 receptors. PLoS One 2012; 7(7): e40187.
       [http://dx.doi.org/10.1371/journal.pone.0040187] [PMID: 22808116]

[9]   Liechti ME, Vollenweider FX. Which neuroreceptors mediate the subjective effects of MDMA in humans? A summary of mechanistic studies. Hum Psychopharmacol 2001; 16(8): 589-98.
       [http://dx.doi.org/10.1002/hup.348] [PMID: 12404538]

[10]   Hysek CM, Fink AE, Simmler LD, Donzelli M, Grouzmann E, Liechti ME. Alpha adrenergic receptors contribute to the acute effects of MDMA in humans. J Clin Psychopharmacol 2013; 33: 658-66.
       [http://dx.doi.org/10.1097/JCP.0b013e3182979d32] [PMID: 23857311]

[11]   Bexis S, Docherty JR. Role of α_{2A}-adrenoceptors in the effects of MDMA on body temperature in the mouse. Br J Pharmacol 2005; 146(1): 1-6.
       [http://dx.doi.org/10.1038/sj.bjp.0706320] [PMID: 16025144]

[12]   Kirkpatrick MG, Francis SM, Lee R, De Wit H, Jacob S. Plasma oxytocin concentrations following MDMA or intranasal oxytocin in humans. Psychoneuroendocrinology 2014; 46: 23-31.
       [http://dx.doi.org/10.1016/j.psyneuen.2014.04.006] [PMID: 24882155]

[13]   Kuypers KPC, Dolder PC, Ramaekers JG, Liechti ME. Multifaceted empathy of healthy volunteers after single doses of MDMA: A pooled sample of placebo-controlled studies. J Psychopharmacol 2017; 31(5): 589-98.
[http://dx.doi.org/10.1177/0269881117699617] [PMID: 28372480]

[14]   Costa G, Caputi FF, Serra M, *et al.* Activation of antioxidant and proteolytic pathways in the nigrostriatal dopaminergic system after 3,4-methylenedioxymethamphetamine administration: Sex-related differences. Front Pharmacol 2021; 12: 713486.
[http://dx.doi.org/10.3389/fphar.2021.713486] [PMID: 34512343]

[15]   Li IH, Ma KH, Weng SJ, Huang SS, Liang CM, Huang YS. Autophagy activation is involved in 3,4-methylenedioxymethamphetamine ('ecstasy') induced neurotoxicity in cultured cortical neurons. PLoS One 2014; 9(12): e116565.
[http://dx.doi.org/10.1371/journal.pone.0116565] [PMID: 25551657]

[16]   Li IH, Ma KH, Kao TJ, *et al.* Involvement of autophagy upregulation in 3,4-methylenedioxymethamphetamine ('ecstasy') induced serotonergic neurotoxicity. Neurotoxicology 2016; 52: 114-26.
[http://dx.doi.org/10.1016/j.neuro.2015.11.009] [PMID: 26610922]

[17]   Mercer LD, Higgins GC, Lau CL, Lawrence AJ, Beart PM. MDMA induced neurotoxicity of serotonin neurons involves autophagy and rilmenidine is protective against its pathobiology. Neurochem Int 2017; 105: 80-90.
[http://dx.doi.org/10.1016/j.neuint.2017.01.010] [PMID: 28122248]

[18]   Lanteri C, Doucet EL, Hernández Vallejo SJ, *et al.* Repeated exposure to MDMA triggers long-term plasticity of noradrenergic and serotonergic neurons. Mol Psychiatry 2014; 19(7): 823-33.
[http://dx.doi.org/10.1038/mp.2013.97] [PMID: 23958955]

[19]   Kindlundh-Högberg AMS, Pickering C, Wicher G, Hobér D, Schiöth HB, Fex Svenningsen Å. MDMA (Ecstasy) decreases the number of neurons and stem cells in embryonic cortical cultures. Cell Mol Neurobiol 2010; 30(1): 13-21.
[http://dx.doi.org/10.1007/s10571-009-9426-y] [PMID: 19543826]

[20]   Bavato F, Stamatakos S, Ohki CMY, *et al.* Brain derived neurotrophic factor protects serotonergic neurons against 3,4-methylenedioxymethamphetamine ("Ecstasy") induced cytoskeletal damage. J Neural Transm 2022; 129(5-6): 703-11.
[http://dx.doi.org/10.1007/s00702-022-02502-8] [PMID: 35420371]

[21]   Ricaurte GA, Martello AL, Katz JL, Martello MB. Lasting effects of (+-)-3-4-methylenedioxymethamphetamine (MDMA) on central serotonergic neurons in nonhuman primates: neurochemical observations. J Pharmacol Exp Ther 1992; 261(2): 616-22.
[PMID: 1374470]

[22]   Federici M, Sebastianelli L, Natoli S, Bernardi G, Mercuri NB. Electrophysiologic changes in ventral midbrain dopaminergic neurons resulting from (+/-) -3,4-methylenedioxymethamphetamine (MDMA-"Ecstasy"). Biol Psychiatry 2007; 62(6): 680-6.
[http://dx.doi.org/10.1016/j.biopsych.2006.11.019] [PMID: 17511969]

[23]   Zakaria FH, Abu Bakar NH, Mohamad N, Ahmad ZAL, Ed SKMJK, Ahmad WANW. The effects of MDMA on brain: An *in Vivo* study in rats. Int J Pharmac Res All Sci 2018; 7(3): 126-37.

[24]   Granado N, O'Shea E, Bove J, Vila M, Colado MI, Moratalla R. Persistent MDMA induced dopaminergic neurotoxicity in the striatum and substantia nigra of mice. J Neurochem 2008; 107(4): 1102-12.
[http://dx.doi.org/10.1111/j.1471-4159.2008.05705.x] [PMID: 18823368]

[25]   Barbosa DJ, Serrat R, Mirra S, *et al.* The mixture of "ecstasy" and its metabolites impairs mitochondrial fusion/fission equilibrium and trafficking in hippocampal neurons, at *in vivo* relevant concentrations. Toxicol Sci 2014; 139(2): 407-20.
[http://dx.doi.org/10.1093/toxsci/kfu042] [PMID: 24595818]

[26]   Selvaraj S, Hoshi R, Bhagwagar Z, *et al.* Brain serotonin transporter binding in former users of MDMA ('ecstasy'). Br J Psychiatry 2009; 194(4): 355-9.
[http://dx.doi.org/10.1192/bjp.bp.108.050344] [PMID: 19336788]

[27]   Buchert R, Thomasius R, Nebeling B, *et al.* Long-term effects of "ecstasy" use on serotonin transporters of the brain investigated by PET. J Nucl Med 2003; 44(3): 375-84.
[PMID: 12621003]

[28]   Dzietko M, Sifringer M, Klaus J, *et al.* Neurotoxic effects of MDMA (ecstasy) on the developing rodent brain. Dev Neurosci 2010; 32(3): 197-207.
[http://dx.doi.org/10.1159/000313473] [PMID: 20616555]

[29]   Aguilar MA, García-Pardo MP, Parrott AC. Of mice and men on MDMA: A translational comparison of the neuropsychobiological effects of 3,4-methylenedioxymethamphetamine ('Ecstasy'). Brain Res 2020; 1727: 146556.
[http://dx.doi.org/10.1016/j.brainres.2019.146556] [PMID: 31734398]

[30]   Karimi S, Jahanshahi M, Golalipour MJ. The effect of MDMA-induced anxiety on neuronal apoptosis in adult male rats' hippocampus. Folia Biol 2014; 60(4): 187-91.
[PMID: 25152052]

[31]   Kermanian F, Seghatoleslam M, Mahakizadeh S. MDMA related neuro-inflammation and adenosine receptors. Neurochem Int 2022 Feb; 153: 105275. Epub 2022 Jan 3.
[http://dx.doi.org/34990730.10.1016/j.neuint.2021.105275.] [PMID: 10366642]

[32]   Liechti M, Saur MR, Gamma A, Hell D, Vollenweider FX. Psychological and physiological effects of MDMA ("Ecstasy") after pretreatment with the 5-HT(2) antagonist ketanserin in healthy humans. Neuropsychopharmacology 2000; 23(4): 396-404.
[http://dx.doi.org/10.1016/S0893-133X(00)00126-3] [PMID: 10989266]

[33]   Schiavone S, Neri M, Maffione A, *et al.* Increased iNOS and nitrosative stress in dopaminergic neurons of MDMA exposed rats. Int J Mol Sci 2019; 20(5): 1242.
[http://dx.doi.org/10.3390/ijms20051242] [PMID: 30871034]

[34]   Torres E, Gutierrez-Lopez MD, Borcel E, *et al.* Evidence that MDMA ('ecstasy') increases cannabinoid CB2 receptor expression in microglial cells: role in the neuroinflammatory response in rat brain. J Neurochem 2010; 113(1): 67-78.
[http://dx.doi.org/10.1111/j.1471-4159.2010.06578.x] [PMID: 20067581]

[35]   Orio L, Llopis N, Torres E, Izco M, O'Shea E, Colado MI. A study on the mechanisms by which minocycline protects against MDMA ('ecstasy')-induced neurotoxicity of 5-HT cortical neurons. Neurotox Res 2010; 18(2): 187-99.
[http://dx.doi.org/10.1007/s12640-009-9120-3] [PMID: 19777321]

[36]   Orio L, O'Shea E, Sanchez V, *et al.* 3,4-Methylenedioxymethamphetamine increases interleukin-1β levels and activates microglia in rat brain: Studies on the relationship with acute hyperthermia and 5-HT depletion. J Neurochem 2004; 89(6): 1445-53.
[http://dx.doi.org/10.1111/j.1471-4159.2004.02443.x] [PMID: 15189347]

[37]   Torres E, Gutierrez-Lopez MD, Mayado A, Rubio A, O'Shea E, Colado MI. Changes in interleukin-1 signal modulators induced by 3,4-methylenedioxymethamphetamine (MDMA): Regulation by CB2 receptors and implications for neurotoxicity. J Neuroinflammation 2011; 8(1): 53.
[http://dx.doi.org/10.1186/1742-2094-8-53] [PMID: 21595923]

[38]   Baumann MH, Rothman RB. Neural and cardiac toxicities associated with 3,4-methylenedioxymethamphetamine (MDMA). Int Rev Neurobiol 2009; 88: 257-96.
[http://dx.doi.org/10.1016/S0074-7742(09)88010-0] [PMID: 19897081]

[39]   McCann UD, Eligulashvili V, Ricaurte GA. (+/-)3,4-Methylenedioxymethamphetamine ('Ecstasy')-induced serotonin neurotoxicity: Clinical studies. Neuropsychobiology 2000; 42(1): 11-6.
[http://dx.doi.org/10.1159/000026665] [PMID: 10867551]

[40] Ricaurte GA, McCann UD, Szabo Z, Scheffel U. Toxicodynamics and long term toxicity of the recreational drug, 3,4-methylenedioxymethamphetamine (MDMA, 'Ecstasy'). Toxicol Lett 2000; 112-113: 143-6.
[http://dx.doi.org/10.1016/S0378-4274(99)00216-7] [PMID: 10720723]

[41] Mueller M, Yuan J, McCann UD, Hatzidimitriou G, Ricaurte GA. Single oral doses of (±) 3,4-methylenedioxymethamphetamine ('Ecstasy') produce lasting serotonergic deficits in non-human primates: Relationship to plasma drug and metabolite concentrations. Int J Neuropsychopharmacol 2013; 16(4): 791-801.
[http://dx.doi.org/10.1017/S1461145712000582] [PMID: 22824226]

[42] McCann UD, Szabo Z, Scheffel U, Dannals RF, Ricaurte GA. Positron emission tomographic evidence of toxic effect of MDMA ("Ecstasy") on brain serotonin neurons in human beings. Lancet 1998; 352(9138): 1433-7.
[http://dx.doi.org/10.1016/S0140-6736(98)04329-3] [PMID: 9807990]

[43] McCann UD, Szabo Z, Vranesic M, et al. Positron emission tomographic studies of brain dopamine and serotonin transporters in abstinent (±)3,4-methylenedioxymethamphetamine ("ecstasy") users: Relationship to cognitive performance. Psychopharmacology 2008; 200(3): 439-50.
[http://dx.doi.org/10.1007/s00213-008-1218-4] [PMID: 18661256]

[44] Erritzoe D, Frokjaer VG, Holst KK, et al. In vivo imaging of cerebral serotonin transporter and serotonin(2A) receptor binding in 3,4-methylenedioxymethamphetamine (MDMA or "ecstasy") and hallucinogen users. Arch Gen Psychiatry 2011; 68(6): 562-76.
[http://dx.doi.org/10.1001/archgenpsychiatry.2011.56] [PMID: 21646575]

CHAPTER 4

LSD (Lysergic Acid Diethylamide)

Abstract: LSD is a potent hallucinogen. It was first synthesised in 1938. It is marketed under numerous names. Ergot, a fungus that develops on rye and grains, is used to make LSD. The effect of LSD is mind-altering, pleasurable, and stimulating. Sometimes, exposure to this drug causes so-called unpleasant experiences, such as "bad trips". It is classified as a Class 1 drug (highly abused) by the Drug Enforcement Agency. Paranoia or psychosis can occur as a negative sequence of taking LSD. Changes in perception, sense of time and space, and mood are reported due to the use of LSD. This medication can be taken orally or through the tongue using tablets, droplets, or blotter paper. LSD is marketed in the streets as blotter paper, thin squares of gelatin, tablet form, liquid sugar cubes, and pure liquid form. Since this is a mind-altering drug, it causes changes in serotonin levels in the brain. LSD affects one's ability to make rational decisions. Speaking with a healthcare professional, talk therapy, and additional medical therapy are options since there is no medication to treat LSD.

Keywords: Brain , Hallucinogen, LSD, Mind-altering drugs, Paranoia, Psychosis, Serotonin levels .

INTRODUCTION

Swiss chemist Albert Hofmann (1938) synthesized LSD [1] working in the Sandoz laboratories in Basel, Switzerland. LSD, the series' 25th molecule, was discovered to be able to be used as an ergot derivative. Hoffman consumed 250ug and experienced a mixture of dizziness, perceptual distortion, confusion, and a tremendous fear of going insane (Hoffman 2009). LD50 values were calculated for the use of LSD in aminals and they were found to differ from species to species. For example, an LD50 value of 50–60 mg/kg was found in mice, whereas for rabbits, it was found to be 0.3 mg/kg [2]. Cutting-edge neuroimaging studies have revealed the disintegration of cortico-striato-thalamo-cortical (CSTC) pathways to lead to a psychedelic state in LSD users [3].

LSD enhances the effect of MDMA, causing an increase in dehydration and heatstroke. A mixture of LSD with MDMA can cause overdose and death. The effects of LSD start decreasing in 24 hrs and may lead to paranoia, depression, and panic; however, MDMA takes a little more time to demonstrate deteriorating effects. LSD and MDMA, if taken together, can cause various symptoms, such as memory problems, decreased appetite, aggression, irritability, insomnia, and trouble concentrating.

In male mice, the effects of LSD on social behaviour using *in vivo* electrophysiology, optogenetics, behavioral paradigms, and molecular biology, were studied. These studies were carried out to understand glutamatergic neurotransmission in the medial prefrontal cortex (mPFC). LSD promoted social behaviour by activating 5-HT_{2A}/AMPA/mTORC1 in excitatory neurotransmission [4]. Epigenetic changes were observed in the rat prefrontal cortex due to LSD administration [5]. LSD interacts with lysosomal cells in neurons, causing behavioural changes (Hendelman 1972). Upon ingestion, LSD is converted to 2-oxo-3-hydroxy-Lsd by liver enzymes. Serotonin receptors and dopamine receptors are the targets for LSD, in particular, all serotonin receptors and D2 receptors [6]. Serotonin 2B receptor subclass was found to be the receptor for binding of LSD to its diethylamide moiety [7]. Up to 20 micrograms of the drug LSD is enough to exert its effects on humans [8]. 5-HT_{2A} receptor is also known to play an essential role in eliciting the effect of LSD [9].

Neurocognitive Effects

LSD decreases the function of attention and cognition [10], and psychomotor functions are also impaired by LSD [11]. Acute anxiety or depression are the main symptoms after the initial use of LSD [Drug Enforcement Administration (2018); D-Lysergic Acid Diethylamide]. LSD initiates the primary process of thinking *via* activation of 5-HT2A receptors [12]. Memory function is affected by LSD use, which especially leads to the impairment of visual memory [13]. The thinking process and intellectual ability are also compromised by LD intake [14]. The effect of LSD on mood changes has been reported to be concentration-dependent. A low dose of LSD (5 mcg) caused anxiety and confusion at 20 mcg [15]. Risk-based decision-making is not affected by LSD, but it affects working memory, executive functions, and cognitive flexibility [16]. The psychosensory responses of LSD are due to the activation of the 5-HT2A receptors and modulation of 5-HT2C and 5-HT1A receptors [9, 17]. LSD could activate various intracellular singling cascades in the male Sprague-Dawley rats' brains; it has been reported to upregulate genes related to cytoskeletal maintenance, glutamate signaling, and synaptic plasticity [18]. LSD affects the signaling of dopamine neurons through the 5-HT1A, D2, and TAAR1 receptors [19]. Hippocampal place cell firing is reduced greatly under the influence of LSD in rats [20]. Tolerance in rats triggered upon LSD treatment due to the reduced serotonin receptor signals in the rat neocortex [21].

5-HT(2A) receptors in the anterior cingulated cortex and medial prefrontal cortex were activated by LSD, confirming the role of these areas in hallucinations [22]. 5-HT(2A) receptor is also involved in the expression of genes involved in schizophrenic hallucinations caused by LSD [23]. 5-HT(2A) receptor is also involved in activating the MAP kinase pathway by upregulating genes, such as C/EBP-beta, MKP-1, and ILAD-1 in the mammalian prefrontal cortex [17]. Apart from that, LSD is known to activate genes that are involved in synaptic plasticity, cytoskeletal architecture, and glutamatergic signaling [18]. 5-HT2 receptor is involved in the binding of LSD and also other hallucinations, as determined by *in vitro* radiolabeling methods [24]. 5-HT(2A) receptor (2AR) binds with all three hallucinogens, such as mescaline, psilocybin, and LSD, suggesting that in the cortex, these 2AR receptors mediate the behavioural responses [25]. Psychotic actions of LSD are mediated by the binding of LSD to the dopaminergic D2R promoter complex [26]. In interneurons of rat piriform cortex, LSD and phenethylamine were also proven to be potent agonists to the 5-HT(2A) receptor (2AR) [27]. Further studies state that the Ca^{2+}/CaM-KII-dependent signal transduction pathway is involved in mediating the NMDA receptor-mediated hallucinogenic effects by LSD [28]. In the monkey brain, selective binding of LSD was found to be the cause of dopamine-mediated agonistic action in producing hallucinogens [29]. LSD was found to be an agonist for dopamine and serotonin receptors in the central nervous system [30]. Increased activation of serotonin receptors due to LSD is mainly responsible for behavioural syndromes in rats exposed to systemic LSD [31]. Some of the behavioural changes associated with LSD were reported to be due to the influence of both 5-HT2A and 5-HT2C receptors [32].

Organotypic cultures of mouse cerebellum were exposed to LSD and it was found to cause endocytosis and changes in the metabolism of the cells [33]. Brain-derived neurotrophic factors were increased in blood plasma by single low doses of LSD (5, 10, and 20 µg) in healthy volunteers [15]. LSD mediated social behaviour through mTOR pathways, which is the mechanistic target of rapamycin complex 1 in the medial prefrontal cortex, in male mice [4]. LSD (75 *µg*, intravenously) in healthy volunteers caused psychosis-like symptoms [34]. In male Sprague-Dawley rats treated with LSD, the temporal phase of behavioural changes was mediated by D2 dopamine receptors [35]. Hippocampal-cortical interaction was suppressed by LSD during active behaviour [20]. LSD was found to increase signal entropy by making a gene expression network; this has also been proven by neuroimaging in the human brain [5]. In a D2 dopamine receptor-mediated fashion, LSD modulated the firing activity of reticular thalamus neurons, thereby altering the state of consciousness in humans [36]. However, LSD is known to increase connectivity from the thalamus to the posterior cingulated cortex *via* serotonin 2A receptor activation [3]. LSD-induced visual imagery was reported to be due to 5-HT2AR activation, leading to the inhibitory processing of neurons between the hippocampus and prefrontal cortex [37 - 40]. Adult male mice exposed to LSD doses (5-160 µg/kg, intraperitoneal) demonstrated disinhibited mediodorsal thalamus relay neurons in a D2 dopamine receptor-mediated fashion upon LSD treatment [36, 46 - 49].

CONCLUSION

LSD enhances the effect of MDMA, causing an increase in dehydration and heatstroke. A mixture of LSD with MDMA can cause overdose and death [40-45]. The effects of LSD start decreasing in 24 hrs and may lead to paranoia, depression, and panic; however, MDMA takes a little more time to demonstrate deteriorating effects. LSD and MDMA, if taken together, can cause various symptoms, such as memory problems, decreased appetite, aggression, irritability, insomnia, and trouble concentrating. LSD decreases the function of attention and cognition [10], and psychomotor functions are also impaired by LSD. Acute anxiety or depression are the main symptoms after the initial use of LSD [Drug Enforcement Administration (2018); D-Lysergic Acid Diethylamide]. LSD initiates the primary process of thinking *via* activation of 5-HT2A receptors. Memory function is affected by LSD use, which especially leads to the impairment of visual memory. The thinking process and intellectual ability are also compromised by LD intake [14]. The effect of LSD on mood changes has been reported to be concentration dependent.

REFERENCES

[1] Stoll A, Hofmann A. Partialsynthese von Alkaloiden vom Typus des Ergobasins. (6. Mitteilung über Mutterkornalkaloide). Helv Chim Acta 1943; 26(3): 944-65.
[http://dx.doi.org/10.1002/hlca.19430260326]

[2] Fang C, Liu JT, Chou SH, Lin CH. Determination of lysergic acid diethylamide (LSD) in mouse blood by capillary electrophoresis/ fluorescence spectroscopy with sweeping techniques in micellar electrokinetic chromatography. Electrophoresis 2003 Mar; 24(6): 1031-7.
[http://dx.doi.org/10.1002/elps.200390119] [PMID: 12658692]

[3] Gregorio DD, Popic J. Lysergic acid diethylamide (LSD) promotes social behavior through mTORC1 in the excitatory neurotransmission. BiolSci 2021; 118(5): e2020705118.

[4] Hofmann A. The discovery of LSD and subsequent investigations on naturally occuring hallucinogens. Discoveries in Biological Psychiatry. Philadelphia, PA: Lippincott 1970.

[5] Jarvik ME, Abramson HA, Hirsch MW. Lysergic acid diethylamide: Effects upon recall and recognition of various stimuli. J Psychol 1955; 39(2): 443-54.
[http://dx.doi.org/10.1080/00223980.1955.9916194]

[6] Abramson HA, Jarvik ME, Kaufman MR, Kornetsky C, Levine A, Wagner M. Lysergic acid diethylamide (LSD-25): I. Physiological and perceptual response. J Psychol 1955; 39(1): 3-60.
[http://dx.doi.org/10.1080/00223980.1955.9916156]

[7] Kornetsky C. Relation of physiological and psychological effects of lysergic acid diethylamide. Arch Neurol Psychiatry 1957; 77(6): 657-8.
[http://dx.doi.org/10.1001/archneurpsyc.1957.02330360115013] [PMID: 13423940]

[8]   Kraehenmann R. LSD increases primary process thinking *via* serotonin 2a receptor activation. Front Pharmacol 2017.

[9]   Abramson HA, Waxenberg SE, Levine A, Kaufman MR, Kornetsky C. Lysergic acid diethylamide (LSD-25): XIII Effect on Bender-Gestalt test performance. J Psychol 1955; 40(2): 341-9.
        [http://dx.doi.org/10.1080/00223980.1955.9712989]

[10]   Isbell H, Fraser HF, Isbell H, Logan CR, Wikler A. Studies on lysergic acid diethylamide (LSD-25). I. Effects in former morphine addicts and development of tolerance during chronic intoxication. AMA Arch Neurol Psychiatry 1956; 76(5): 468-78.
        [http://dx.doi.org/10.1001/archneurpsyc.1956.02330290012002] [PMID: 13371962]

[11]   Hutten NRPW, Mason NL, Dolder PC, *et al.* Mood and cognition after administration of low LSD doses in healthy volunteers: A placebo controlled dose-effect finding study. Eur Neuropsychopharmacol 2020; 41(December): 81-91.
        [http://dx.doi.org/10.1016/j.euroneuro.2020.10.002] [PMID: 33082016]

[12]   Pokorny T, Duerler P, Seifritz E, Vollenweider FX, Preller KH. LSD acutely impairs working memory, executive functions, and cognitive flexibility, but not risk-based decision-making. Psychol Med 2020; 50(13): 2255-64.
        [http://dx.doi.org/10.1017/S0033291719002393] [PMID: 31500679]

[13]   Nichols DE. Hallucinogens. Pharmacol Ther 2004; 101(2): 131-81.
        [http://dx.doi.org/10.1016/j.pharmthera.2003.11.002] [PMID: 14761703]

[14]   Vollenweider F, Leenders KL, Scharfetter C, Maguire P, Stadelmann O, Angst J. Positron emission tomography and fluorodeoxyglucose studies of metabolic hyperfrontality and psychopathology in the psilocybin model of psychosis. Neuropsychopharmacology 1997; 16(5): 357-72.
        [http://dx.doi.org/10.1016/S0893-133X(96)00246-1] [PMID: 9109107]

[15]   Hofmann A. LSD My Problem Child. Santa Cruz: Multidisciplinary Association for Psychedelic Studies 2009.

[16]   Nichols C, Sanders-Bush E. A single dose of lysergic acid diethylamide influences gene expression patterns within the mammalian brain. Neuropsychopharmacology 2002; 26(5): 634-42.
        [http://dx.doi.org/10.1016/S0893-133X(01)00405-5] [PMID: 11927188]

[17]   De Gregorio D, Posa L, Ochoa-Sanchez R, *et al.* The hallucinogen d-lysergic diethylamide (LSD) decreases dopamine firing activity through 5-HT1A, D2 and TAAR1 receptors. Pharmacol Res 2016; 113((Pt A)): 81-91.
        [http://dx.doi.org/10.1016/j.phrs.2016.08.022]

[18]   Herr KA, Baker LE. Re-evaluation of the discriminative stimulus effects of lysergic acid diethylamide with male and female Sprague-Dawley rats. Behav Pharmacol 2020 Dec; 31(8): 776-786.
        [http://dx.doi.org/10.1097/FBP.0000000000000589] [PMID: 32960851]

[19]   Hendelman WJ. A morphologic study of the effects of LSD on neurons in cultures of cerebelum. J Neuropathol Exp Neurol 1972; 31(3): 411-32.
        [http://dx.doi.org/10.1097/00005072-197207000-00002] [PMID: 5055392]

[20]   Dolder PC, Schmid Y, Haschke M, Rentsch KM, Liechti ME. Pharmacokinetics and concentration-effect relationship of oral LSD in humans. Int J Neuropsychopharmacol 2016; 19(1): pyv072.
        [http://dx.doi.org/10.1093/ijnp/pyv072] [PMID: 26108222]

[21]   Wacker D, Wang S, McCorvy JD, *et al.* Crystal structure of an lsd-bound human serotonin receptor. Cell 2017; 168(3): 377-389.e12.
        [http://dx.doi.org/10.1016/j.cell.2016.12.033] [PMID: 28129538]

[22]   Greiner T, Burch NR, Edelberg R. Psychopathology and psychophysiology of minimal LSD-25 dosage; a preliminary dosage-response spectrum. AMA Arch Neurol Psychiatry 1958; 79(2): 208-10.
        [http://dx.doi.org/10.1001/archneurpsyc.1958.02340020088016] [PMID: 13497365]

[23]   Vollenweider FX, Preller KH. Psychedelic drugs: neurobiology and potential for treatment of psychiatric disorders. Nat Rev Neurosci 2020; 21(11): 611-24.
[http://dx.doi.org/10.1038/s41583-020-0367-2] [PMID: 32929261]

[24]   Glennon RA, Titeler M, McKenney JD. Evidence for 5-HT2 involvement in the mechanism of action of hallucinogenic agents. Life Sci 1984 Dec 17; 35(25): 2505-11.
[http://dx.doi.org/10.1016/0024-3205(84)90436-3] [PMID: 6513725]

[25]   Gresch PJ, Smith RL, Barrett RJ, Sanders-Bush E. Behavioral tolerance to lysergic acid diethylamide is associated with reduced serotonin-2A receptor signaling in rat cortex. Neuropsychopharmacology 2005; 30(9): 1693-702.
[http://dx.doi.org/10.1038/sj.npp.1300711] [PMID: 15756304]

[26]   Gresch PJ, Strickland LV, Sanders-Bush E. Lysergic acid diethylamide-induced Fos expression in rat brain: Role of serotonin-2A receptors. Neuroscience 2002; 114(3): 707-13.
[http://dx.doi.org/10.1016/S0306-4522(02)00349-4] [PMID: 12220572]

[27]   Nichols CD, Garcia EE, Sanders-Bush E. Dynamic changes in prefrontal cortex gene expression following lysergic acid diethylamide administration. Brain Res Mol Brain Res 2003; 111(1-2): 182-8.
[http://dx.doi.org/10.1016/S0169-328X(03)00029-9] [PMID: 12654518]

[28]   Nichols CD, Sanders-Bush E. Molecular genetic responses to lysergic acid diethylamide include transcriptional activation of MAP kinase phosphatase-1, C/EBP-β and ILAD-1, a novel gene with homology to arrestins. J Neurochem 2004; 90(3): 576-84.
[http://dx.doi.org/10.1111/j.1471-4159.2004.02515.x] [PMID: 15255935]

[29]   Frederick DL, Gillam MP, Lensing S, Paule MG. Acute effects of LSD on rhesus monkey operant test battery performance. Pharmacol Biochem Behav 1997 Aug; 57(4): 633-41.
[http://dx.doi.org/10.1016/s0091-3057(96)00469-8] [PMID: 9258988]

[30]   Titeler M, Lyon RA, Glennon RA. Radioligand binding evidence implicates the brain 5-HT2 receptor as a site of action for LSD and phenylisopropylamine hallucinogens. Psychopharmacology 1988; 94(2): 213-6.
[http://dx.doi.org/10.1007/BF00176847] [PMID: 3127847]

[31]   De Gregorio D, Inserra A, Enns JP, *et al.* Repeated lysergic acid diethylamide (LSD) reverses stress-induced anxiety-like behavior, cortical synaptogenesis deficits and serotonergic neurotransmission decline. Neuropsychopharmacology 2022 May; 47(6): 1188-1198. Epub 2022 Mar 17.
[http://dx.doi.org/10.1038/s41386-022-01301-9] [PMID: 35301424] [PMCID: PMC9018770]

[32]   Hutten NRPW, Mason NL, Dolder PC, *et al.* Low doses of lsd acutely increase bdnf blood plasma levels in healthy volunteers. ACS Pharmacol Transl Sci 2021; 4(2): 461-6.
[http://dx.doi.org/10.1021/acsptsci.0c00099] [PMID: 33860175]

[33]   De Gregorio D, Popic J, Enns JP, *et al.* Lysergic acid diethylamide (LSD) promotes social behavior through mTORC1 in the excitatory neurotransmission. Proc Natl Acad Sci 2021; 118(5): e2020705118.
[http://dx.doi.org/10.1073/pnas.2020705118] [PMID: 33495318]

[34]   González-Maeso J, Weisstaub NV, Zhou M, *et al.* Hallucinogens recruit specific cortical 5-HT(2A) receptor-mediated signaling pathways to affect behavior. Neuron 2007; 53(3): 439-52.
[http://dx.doi.org/10.1016/j.neuron.2007.01.008] [PMID: 17270739]

[35]   Borroto-Escuela DO, Romero-Fernandez W, Narvaez M, Oflijan J, Agnati LF, Fuxe K. Hallucinogenic 5-HT2AR agonists LSD and DOI enhance dopamine D2R protomer recognition and signaling of D2-5-HT2A heteroreceptor complexes. Biochem Biophys Res Commun 2014; 443(1): 278-84.
[http://dx.doi.org/10.1016/j.bbrc.2013.11.104] [PMID: 24309097]

[36]   Carhart-Harris RL, Kaelen M, Bolstridge M, *et al.* The paradoxical psychological effects of lysergic acid diethylamide (LSD). Psychol Med 2016; 46(7): 1379-90.

[http://dx.doi.org/10.1017/S0033291715002901] [PMID: 26847689]

[37] Marek GJ, Aghajanian GK. LSD and the phenethylamine hallucinogen DOI are potent partial agonists at 5-HT2A receptors on interneurons in rat piriform cortex. J Pharmacol Exp Ther 1996; 278(3): 1373-82.
[PMID: 8819525]

[38] Arvanov VL, Liang X, Russo A, Wang RY. LSD and DOB: interaction with 5-HT $_{2A}$ receptors to inhibit NMDA receptor-mediated transmission in the rat prefrontal cortex. Eur J Neurosci 1999; 11(9): 3064-72.
[http://dx.doi.org/10.1046/j.1460-9568.1999.00726.x] [PMID: 10510170]

[39] Ahn HS, Makman MH. Interaction of LSD and other hallucinogens with dopamine-sensitive adenylate cyclase in primate brain: Regional differences. Brain Res 1979; 162(1): 77-88.
[http://dx.doi.org/10.1016/0006-8993(79)90757-1] [PMID: 104775]

[40] Von Hungen K, Roberts S, Hill DF. Interactions between lysergic acid diethylamide and dopamine-sensitive adenylate cyclase systems in rat brain. Brain Res 1975; 94(1): 57-66.
[http://dx.doi.org/10.1016/0006-8993(75)90876-8] [PMID: 238721]

[41] Trulson ME, Ross CA, Jacobs BL. Behavioral evidence for the stimulation of CNS serotonin receptors by high doses of LSD. Psychopharmacol Commun 1976; 2(2): 149-64.
[PMID: 136010]

[42] Krebs-Thomson K, Paulus MP, Geyer MA. Effects of hallucinogens on locomotor and investigatory activity and patterns: influence of 5-HT2A and 5-HT2C receptors. Neuropsychopharmacology 1998; 18(5): 339-51.
[http://dx.doi.org/10.1016/S0893-133X(97)00164-4] [PMID: 9536447]

[43] Marona-Lewicka D, Thisted RA, Nichols DE. Distinct temporal phases in the behavioral pharmacology of LSD: dopamine D2 receptor-mediated effects in the rat and implications for psychosis. Psychopharmacology 2005; 180(3): 427-35.
[http://dx.doi.org/10.1007/s00213-005-2183-9] [PMID: 15723230]

[44] Domenico C, Haggerty D, Mou X, Ji D. LSD degrades hippocampal spatial representations and suppresses hippocampal-visual cortical interactions. Cell Rep 2021; 36(11): 109714.
[http://dx.doi.org/10.1016/j.celrep.2021.109714] [PMID: 34525364]

[45] Savino A, Nichols CD. Lysergic acid diethylamide induces increased signalling entropy in rats' prefrontal cortex. J Neurochem 2022; 162(1): 9-23.
[http://dx.doi.org/10.1111/jnc.15534] [PMID: 34729786]

[46] De Gregorio D, Popic J, Enns JP, et al. Lysergic acid diethylamide (LSD) promotes social behavior through mTORC1 in the excitatory neurotransmission. Proc Natl Acad Sci U S A 2021; 35(4): 469-82.
[http://dx.doi.org/10.1073/pnas.2020705118] [PMID: 33495318] [PMID: PMC7865169]

[47] Preller KH, Razi A, Zeidman P, Stämpfli P, Friston KJ, Vollenweider FX. Effective connectivity changes in LSD-induced altered states of consciousness in humans. Proc Natl Acad Sci USA 2019; 116(7): 2743-8.
[http://dx.doi.org/10.1073/pnas.1815129116] [PMID: 30692255]

[48] Schmidt A, Müller F, Lenz C, et al. Acute LSD effects on response inhibition neural networks. Psychol Med 2018; 48(9): 1464-73.
[http://dx.doi.org/10.1017/S0033291717002914] [PMID: 28967351]

[49] Inserra A, De Gregorio D, Rezai T, Lopez-Canul MG, Comai S, Gobbi G. Lysergic acid diethylamide differentially modulates the reticular thalamus, mediodorsal thalamus, and infralimbic prefrontal cortex: An in vivo electrophysiology study in male mice. J Psychopharmacol 2021; 35(4): 469-82.
[http://dx.doi.org/10.1177/0269881121991569] [PMID: 33645311]

CHAPTER 5

Methamphetamine

Abstract: Methamphetamine (METH) is a highly addictive stimulant that affects the central nervous system. It is a widely abused psychostimulant. Monoaminergic neurotransmitter terminals are affected by METH intake. METH structure is very similar to amphetamine, a drug used to treat attention-deficit hyperactivity disorder (ADHD). METH is taken in various modes, such as smoking, swallowing, snorting, injecting powder, *etc*. Dopamine levels, serotonin levels, and norepinephrine levels are increased due to METH uptake, leading to extremely strong euphoric effects. This dopamine surge causes the brain to repeatedly take the drug and is responsible for addiction. As a short-term effect, METH causes increased wakefulness and physical activity, decreased appetite, faster breathing, rapid and/or irregular heartbeat, increased blood pressure, and body temperature. METH overdose causes hyperthermia and convolution, which can lead to death if not treated. METH also causes irreversible brain damage. Amphetamine psychosis, dementia-like symptoms, increased anti-social behaviour, and increased susceptibility to neurodegenerative diseases are the long-term neurological effects of METH use.

Keywords: Amphetamine psychosis, Attention-deficit hyperactivity disorder (ADHD), Dopamine levels, Serotonin levels.

INTRODUCTION

Crystal METH is a more powerful and harmful addictive [1] than amphetamine. Chronic psychosis is the most problematic psychiatric issue during MEH usage [1 - 4]. It has been observed that crystal METH abuse causes disturbances to striatal volume [5]. Diffuse white matter alterations were seen in the temporal, frontal/parietal and occipital lobes in METH users [6]. Dopamine transporter levels were increased **72** hours after withdrawal in rats during methamphetamine self-administration **(2)**. METH stays in the brain for a significantly longer time than cocaine. Due to this reason, the drug has long-lasting behavioural and neurotoxic effects. Moreover, it causes brain damage, heart damage, liver and kidney damage, intense itching, severe tooth decay, and tooth loss. Irritability, anxiety, psychosis, and mood disturbance are the common psychiatric symptoms with METH usage [7]. In Australia, it was found that the prevalence of METH usage is 11 p ercent higher [8]. There is also a dose-related increase in violent

Jayalakshmi Krishnan

behaviour during periods of methamphetamine use [9]. Up to 12 hrs, crystal METH produces euphoria in the brain, causing a high risk of physical and psychological disturbance. Some other drugs are also combined with METH to create a stronger effect. Such drugs are alcohol, morphine, and xanax. Microglia activation is one of the phenomena that lead to METH neurotoxicity [7, 10]. Neurotoxicity of METH causes astrogliosis, tyrosine hydroxylase (TH), metabolite dysregulation and imbalances, *etc* [11 - 13]. In a study, it was found that μ-opioid receptor knockout mice were highly insensitive to METH challenge as they demonstrated a change in dopamine receptor ligand binding activity [14]. Dopamine receptor 3 knockout mice revealed that it sensitises the brain to behavioural changes and gene expression [15]. The key differences between METH and amphetamines are as follows: amphetamines are prescription drugs that are prescribed by doctors. They fall into the category of stimulant drugs and are used in medical science for treating some diseases. However, METH is an illegal stimulant and street drug with no medical use. Both these drugs are classified as Schedule II drugs as they are susceptible to abuse and addiction.

Amphetamines and METH

Amphetamines are used for both therapeutic and recreational purposes, so doctors must educate patients on the potential toxicity of therapeutic and recreational amphetamine use. There are two forms of amphetamines: one is a slow-acting oral form of amphetamine (20–60 minutes), and the other is a fast-acting oral form (seconds to minutes) of amphetamine. Slow-onset form is given for medical purposes, but fast-acting forms are taken under unsupervised medical use [16].

For ADHD children, the typical dose of amphetamine is 20–25 mg. Neurodegeneration and cognitive decline are noted due to the continuous use of METH [17, 18]. It has been observed that methamphetamine users develop Parkinson's disease more frequently than non-users [19]. METH-induced neurotoxicity causes oxidative stress, activation of transcription factors, excitotoxicity, and DNA damage. Not only these, but blood-brain barrier breakdown, microglia activation, and overactivation of various apoptotic pathways are also noted [13, 20]. METH-induced neuroinflammation leads to neuropathology. Catecholamine neurotransmission has been affected due to METH use [21].

Impaired Neurogenesis

It has been observed that dentate gyrus neurogenesis is impaired in METH users, followed by cognitive decline [22]. Moreover, dentate gyrus stem cell renewal is also decreased in METH users. In animal studies, it was found that METH causes neurodegeneration in the dopaminergic nerve terminal, thus reducing the expression of tyrosine hydroxylase [23, 24]. Furthermore, hippocampal neurogenesis was found to be impaired by METH addiction in mice [25].

Dopamine and METH

It has been observed that dopamine and vesicular monoamine transporters are decreased in the dopamine neuron membrane by methamphetamine [26, 27]. Dopamine and serotonin are the two neurotransmitters that are released by acute METH exposure [28]. METH has various effects on dopamine: 1) competing with dopamine for binding sites, 2) disrupting the storage of dopamine in the vesicles, and 3) triggering dopamine efflux [24, 29, 30]. The specific mechanism of METH dysregulation by dopamine is due to its binding to the SIGMA receptor, which lies in the membrane of the endoplasmic reticulum, which is also a chaperone [31]. METH stays in the brain for a longer time; hence, it exerts its stimulant properties for a long time [32]. METH particularly affects the nigrostriatal pathway, not affecting the mesolimbic pathway [33]. Interestingly, METH is known to activate Toll-like receptor 4, leading to high dopamine levels in the extracellular space, thus paving the way to finding a cure for drug abuse [34].

METH impairs the Ubiquitin Proteasome System (UPS) through dopamine action [35]. Through activating and channelizing intracellular calcium levels, METH causes the release of dopamine [36]. METH users can develop neurotoxic effects, which may be reflected as edema, stroke, haemorrhage, hallucinations, and paranoia. Its withdrawal produces effects associated with anxiety and extreme cravings for the drug. METH users can develop shortness of breath and increased respiration, ultimately leading to heart attack or stroke.

Oxidative stress is induced by METH in the mitochondria of substantia nigra pars compacta dopaminergic neurons [37]. Chronic METH-induced neurodegeneration is due to mitochondrial stress mediated by calcium channels and monoamine oxidase [38]. Chronic METH intake also induces mitochondrial stress in substantia nigra pars compacta (SNc) axons [39]. Tyrosine hydroxylase mRNA levels were altered in rats due to repeated METH administration [40]. METH administration caused long-lasting changes in dopamine neurons of the nigrostriatal pathway [41]. In male mice, METH treatment caused apoptosis in striatal neurons, such as parvalbumin neurons and cholinergic interneurons [42]. Similar observations were made in the striatum that METH intake damaged striatal neurons and dopaminergic terminals [43]. Furthermore, in somatostatin interneurons, METH caused the trafficking of neurokinin 1 receptors, leading to anti-apoptotic activity [44]. Neuronal nitric oxide synthases were induced by METH, as evidenced by 3-nitrotyrosine in the striatum [45].

METH causes an increase in glial fibrillary acidic protein in NRF2+/+ brain mediated by nuclear factor E2 related factor 2 [46]. METH abuse is related to the loss of dopaminergic neurons, and this is due to the mitochondrial oxidative stress created in the neurons [47]. It is clearly known that METH causes apoptosis and autophagy in neurons [48]. In the hippocampal neuron cell line HT-2, METH induces endoplasmic reticulum stress causing neuronal death [49]. SVGA astrocytes, upon treatment with METH, underwent type 1-programmed cell death, which is due to stress mediated by the endoplasmic reticulum [50]. Long-lasting memory deficits were caused by METH exposure in mice [51]. METH-induced neuronal damage was found to be due to neuroinflammation and neurotoxic effects of the drug, causing DNA damage and excitotoxicity [52]. In male Wistar rats, METH caused spatial memory impairment [53]. It also caused astrogliosis and apoptosis in hippocampal neurons in male Wistar rats [54] and neuroinflammation in hippocampal neurons [55]. Marked changes were observed in METH-treated astrocytes and neural progenitor cells in terms of increased transcription factors of cytokines and inflammasomes [56]. Blood-brain barrier damage and neuroinflammation occur in astrocytes and increased expression of various proinflammatory cytokines [57].

The transfer of alpha-synuclein from neurons to astrocytes *via* exosomes causes inflammatory responses [58]. In primary cultured neuronal cells and mice brains, METH increased the levels of total Tau and pSer396 Tau protein [59]. Alpha-synuclein levels were found to be accumulated in METH-exposed cell lines; the mechanism for this could be based on polyubiquitin and α-syn expression in cell lines [58]. Using a chronic METH mouse model, it was proven that alpha-synuclein is involved in METH-induced toxicity. Alpha-synuclein also causes damage or myelin loss in mice [60]. METH induces toxicity through CCAAT-enhancer binding protein β (C/EBPβ) by the DDIT4/TSC2/mTOR signalling pathway in neurons. This complex mechanism causes autophagy and apoptosis in neuronal cells [61]. In HIV-positive METH abusers, it was found that METH causes oxidative stress. Using the human brain microvascular endothelial cell line hCMEC/D3, it was revealed that blood-brain barrier damage occurs through the transient receptor potential melastatin 2 (TRPM2) channel during METH treatment [62]. METH attacks the blood-brain barrier proteins and causes damage to the barrier by making permeability to many molecules [63]. Tight junction proteins and matrix metalloproteinase are involved in METH-induced blood-brain barrier dysfunction in the hippocampus [22]. Rho/ROCK dependent pathway and F actin cytoskeleton are involved in the METH-induced blood-brain permeability [64]. METH causes neurovascular dysfunction, neuroinflammatory events, and blood-brain barrier damage in neurons [65]. In a study on rats pre-exposed to chronic METH, it was found that the blood-brain barrier breakage is due to the activation of cyclooxygenase [66, 67]. Vesicular monoamine transporter 2

(VMAT-2) is involved in METH-induced toxicity, and METH causes hyperthermia in rats exposed to both stress and METH [68]. In striatal synaptosomes exposed to METH, rapid oxidation and a decrease in vesicular monoamine transporter 2 were observed [69 - 71].

CONCLUSION

Crystal METH is a more powerful and harmful addictive than amphetamine. Chronic psychosis is the most problematic psychiatric issue during MEH usage. It has been observed that crystal METH abuse causes disturbances to striatal volume. Diffuse white matter alterations were seen in the temporal, frontal/parietal and occipital lobes in METH users. Dopamine transporter levels were increased 72 hours after withdrawal in rats during methamphetamine self-administration. METH stays in the brain for a significantly longer time than cocaine. Due to this reason, the drug has long-lasting behavioural and neurotoxic effects. Moreover, it causes brain damage, heart damage, liver and kidney damage, intense itching, severe tooth decay, and tooth loss. Irritability, anxiety, psychosis, and mood disturbance are the common psychiatric symptoms with METH usage. It has been observed that dopamine and vesicular monoamine transporters are decreased in the dopamine neuron membrane by methamphetamine. Dopamine and serotonin are the two neurotransmitters that are released by acute METH exposure.

REFERENCES

[1]   Schifano F, Corkery JM, Cuffolo G. Smokable ("ice", "crystal meth") and non smokable amphetamine-type stimulants: Clinical pharmacological and epidemiological issues, with special reference to the UK. Ann Ist Super Sanita 2007; 43(1): 110-5.
[PMID: 17536161]

[2]   D'Arcy C, Luevano JE, Miranda-Arango M, *et al.* Extended access to methamphetamine self-administration up regulates dopamine transporter levels 72 hours after withdrawal in rats. Behav Brain Res 2016; 296: 125-8.
[http://dx.doi.org/10.1016/j.bbr.2015.09.010] [PMID: 26367473]

[3]   Chiadmi F, Schlatter J. Crystal meth : Une forme de méthamphétamine. Presse Med 2009; 38(1): 63-7.
[http://dx.doi.org/10.1016/j.lpm.2008.07.017] [PMID: 18976881]

[4]   Nelson ME, Bryant SM, Aks SE. Emerging drugs of abuse. Emerg Med Clin North Am 2014; 32(1): 1-28.
[http://dx.doi.org/10.1016/j.emc.2013.09.001] [PMID: 24275167]

[5]   Möbius C, Kustermann A, Struffert T, Kornhuber J, Müller HH. c-MRI findings after crystal meth abuse. J Addict Med 2014; 8(5): 384-5.
[http://dx.doi.org/10.1097/ADM.0000000000000051] [PMID: 25026102]

[6]   Zweben JE, Cohen JB, Christian D, *et al.* Psychiatric symptoms in methamphetamine users. Am J Addict 2004; 13(2): 181-90.
[http://dx.doi.org/10.1080/10550490490436055] [PMID: 15204668]

[7]   McKetin R, McLaren J, Lubman DI, Hides L. The prevalence of psychotic symptoms among methamphetamine users. Addiction 2006; 101(10): 1473-8.

[http://dx.doi.org/10.1111/j.1360-0443.2006.01496.x] [PMID: 16968349]

[8] McKetin R, Lubman DI, Najman JM, Dawe S, Butterworth P, Baker AL. Does methamphetamine use increase violent behaviour? Evidence from a prospective longitudinal study. Addiction 2014; 109(5): 798-806.
[http://dx.doi.org/10.1111/add.12474] [PMID: 24400972]

[9] Perez-Reyes M, White WR, McDonald SA, Hill JM, Jeffcoat AR, Cook CE. Clinical effects of methamphetamine vapor inhalation. Life Sci 1991; 49(13): 953-9.
[http://dx.doi.org/10.1016/0024-3205(91)90078-P] [PMID: 1886456]

[10] Dean AC, Groman SM, Morales AM, London ED. An evaluation of the evidence that methamphetamine abuse causes cognitive decline in humans. Neuropsychopharmacology 2013; 38(2): 259-74.
[http://dx.doi.org/10.1038/npp.2012.179] [PMID: 22948978]

[11] Rusyniak DE. Neurologic manifestations of chronic methamphetamine abuse. Neurol Clin 2011; 29(3): 641-55.
[http://dx.doi.org/10.1016/j.ncl.2011.05.004] [PMID: 21803215]

[12] Granado N, Ares-Santos S, Moratalla R. Methamphetamine and Parkinson's disease. Parkinsons Dis 2013; 2013: 1-10.
[http://dx.doi.org/10.1155/2013/308052] [PMID: 23476887]

[13] Cadet JL, Krasnova IN. Molecular bases of methamphetamine induced neurodegeneration. Int Rev Neurobiol 2009; 88: 101-19.
[http://dx.doi.org/10.1016/S0074-7742(09)88005-7] [PMID: 19897076]

[14] Meredith CW, Jaffe C, Ang-Lee K, Saxon AJ. Implications of chronic methamphetamine use: A literature review. Harv Rev Psychiatry 2005; 13(3): 141-54.
[http://dx.doi.org/10.1080/10673220591003605] [PMID: 16020027]

[15] Barbosa D J, Capela J P, Feio-Azevedo R, Teixeira-Gomes A, Bastos M de L, Carvalho F. Mitochondria: Key players in the neurotoxic effects of amphetamines. Arch Toxicol 2015; 89(10): 1695-725.
[http://dx.doi.org/10.1007/s00204-015-1478-9]

[16] Schmidt CJ, Ritter JK, Sonsalla PK, Hanson GR, Gibb JW. Role of dopamine in the neurotoxic effects of methamphetamine. J Pharmacol Exp Ther 1985; 233(3): 539-44.
[PMID: 2409267]

[17] Brown JM, Hanson GR, Fleckenstein AE. Methamphetamine rapidly decreases vesicular dopamine uptake. J Neurochem 2000; 74(5): 2221-3.
[http://dx.doi.org/10.1046/j.1471-4159.2000.0742221.x] [PMID: 10800970]

[18] Baucum AJ II, Rau KS, Riddle EL, Hanson GR, Fleckenstein AE. Methamphetamine increases dopamine transporter higher molecular weight complex formation *via* a dopamine- and hyperthermia-associated mechanism. J Neurosci 2004; 24(13): 3436-43.
[http://dx.doi.org/10.1523/JNEUROSCI.0387-04.2004] [PMID: 15056723]

[19] Halpin LE, Collins SA, Yamamoto BK. Neurotoxicity of methamphetamine and 3,4-methylenedioxymethamphetamine. Life Sci 2014; 97(1): 37-44.
[http://dx.doi.org/10.1016/j.lfs.2013.07.014] [PMID: 23892199]

[20] LaVoie MJ, Card JP, Hastings TG. Microglial activation precedes dopamine terminal pathology in methamphetamine induced neurotoxicity. Exp Neurol 2004; 187(1): 47-57.
[http://dx.doi.org/10.1016/j.expneurol.2004.01.010] [PMID: 15081587]

[21] Loftis JM, Janowsky A. Neuroimmune basis of methamphetamine toxicity. Int Rev Neurobiol 2014; 118: 165-97.
[http://dx.doi.org/10.1016/B978-0-12-801284-0.00007-5] [PMID: 25175865]

[22] Jones SR, Gainetdinov RR, Wightman RM, Caron MG. Mechanisms of amphetamine action revealed

in mice lacking the dopamine transporter. J Neurosci 1998; 18(6): 1979-86.
[http://dx.doi.org/10.1523/JNEUROSCI.18-06-01979.1998] [PMID: 9482784]

[23]   Schmitz Y, Lee CJ, Schmauss C, Gonon F, Sulzer D. Amphetamine distorts stimulation-dependent dopamine overflow: Effects on D2 autoreceptors, transporters, and synaptic vesicle stores. J Neurosci 2001; 21(16): 5916-24.
[http://dx.doi.org/10.1523/JNEUROSCI.21-16-05916.2001] [PMID: 11487614]

[24]   Goodwin JS, Larson GA, Swant J, *et al.* Amphetamine and methamphetamine differentially affect dopamine transporters *in vitro* and *in vivo*. J Biol Chem 2009; 284(5): 2978-89.
[http://dx.doi.org/10.1074/jbc.M805298200] [PMID: 19047053]

[25]   O'Callaghan JP, Miller DB. Neurotoxicity profiles of substituted amphetamines in the C57BL/6J mouse. J Pharmacol Exp Ther 1994; 270(2): 741-51.
[PMID: 8071867]

[26]   Fantegrossi WE, Ciullo JR, Wakabayashi KT, De La Garza R II, Traynor JR, Woods JH. A comparison of the physiological, behavioral, neurochemical and microglial effects of methamphetamine and 3,4-methylenedioxymethamphetamine in the mouse. Neuroscience 2008; 151(2): 533-43.
[http://dx.doi.org/10.1016/j.neuroscience.2007.11.007] [PMID: 18082974]

[27]   Krasnova IN, Justinova Z, Ladenheim B, *et al.* Methamphetamine self-administration is associated with persistent biochemical alterations in striatal and cortical dopaminergic terminals in the rat. PLoS One 2010; 5(1): e8790.
[http://dx.doi.org/10.1371/journal.pone.0008790] [PMID: 20098750]

[28]   Hedges DM, Obray JD, Yorgason JT, *et al.* Methamphetamine induces dopamine release in the nucleus accumbens through a sigma receptor mediated pathway. Neuropsychopharmacology 2018; 43(6): 1405-14.
[http://dx.doi.org/10.1038/npp.2017.291] [PMID: 29185481]

[29]   Chiu VM, Schenk JO. Mechanism of action of methamphetamine within the catecholamine and serotonin areas of the central nervous system. Curr Drug Abuse Rev 2012; 5(3): 227-42.
[http://dx.doi.org/10.2174/1874473711205030227] [PMID: 22998621]

[30]   Granado N, Ares-Santos S, O'Shea E, Vicario-Abejón C, Colado MI, Moratalla R. Selective vulnerability in striosomes and in the nigrostriatal dopaminergic pathway after methamphetamine administration : Early loss of TH in striosomes after methamphetamine. Neurotox Res 2010; 18(1): 48-58.
[http://dx.doi.org/10.1007/s12640-009-9106-1] [PMID: 19760475]

[31]   Wang X, Northcutt AL, Cochran TA, *et al.* Methamphetamine activates toll-like receptor 4 to induce central immune signaling within the ventral tegmental area and contributes to extracellular dopamine increase in the nucleus accumbens shell. ACS Chem Neurosci 2019; 10(8): 3622-34.
[http://dx.doi.org/10.1021/acschemneuro.9b00225] [PMID: 31282647]

[32]   Limanaqi F, Biagioni F, Busceti CL, Ryskalin L, Fornai F. The effects of proteasome on baseline and methamphetamine dependent dopamine transmission. Neurosci Biobehav Rev 2019; 102: 308-17.
[http://dx.doi.org/10.1016/j.neubiorev.2019.05.008] [PMID: 31095962]

[33]   Yorgason JT, Hedges DM, Obray JD, *et al.* Methamphetamine increases dopamine release in the nucleus accumbens through calcium dependent processes. Psychopharmacology 2020; 237(5): 1317-30.
[http://dx.doi.org/10.1007/s00213-020-05459-2] [PMID: 31965252]

[34]   Park SW, Shen X, Tien LT, Roman R, Ma T. Methamphetamine-induced changes in the striatal dopamine pathway in μ-opioid receptor knockout mice. J Biomed Sci 2011; 18(1): 83.
[http://dx.doi.org/10.1186/1423-0127-18-83] [PMID: 22074218]

[35]   Su H, Wang X, Bai J, *et al.* Role of dopamine D3 receptors in methamphetamine-induced behavioural sensitization and the characterization of dopamine receptors (D1R–D5R) gene expression in the brain.

Folia Neuropathol 2022; 60(1): 105-13.
[http://dx.doi.org/10.5114/fn.2022.114021] [PMID: 35359150]

[36]   Graves SM, Schwarzschild SE, Tai RA, Chen Y, Surmeier DJ. Mitochondrial oxidant stress mediates methamphetamine neurotoxicity in substantia nigra dopaminergic neurons. Neurobiol Dis 2021; 156: 105409.
[http://dx.doi.org/10.1016/j.nbd.2021.105409] [PMID: 34082123]

[37]   Du Y, Choi S, Pilski A, Graves SM. Differential vulnerability of locus coeruleus and dorsal raphe neurons to chronic methamphetamine induced degeneration. Front Cell Neurosci 2022; 16: 949923.
[http://dx.doi.org/10.3389/fncel.2022.949923] [PMID: 35936499]

[38]   Du Y, Lee YB, Graves SM. Chronic methamphetamine induced neurodegeneration: Differential vulnerability of ventral tegmental area and substantia nigra pars compacta dopamine neurons. Neuropharmacology 2021; 200: 108817.
[http://dx.doi.org/10.1016/j.neuropharm.2021.108817] [PMID: 34610287]

[39]   Zhang Y, Angulo JA. Contrasting effects of repeated treatment vs. withdrawal of methamphetamine on tyrosine hydroxylase messenger RNA levels in the ventral tegmental area and substantia nigra zona compacta of the rat brain. Synapse 1996; 24(3): 218-23.
[http://dx.doi.org/10.1002/(SICI)1098-2396(199611)24:3<218::AID-SYN3>3.0.CO;2-H]   [PMID: 8923661]

[40]   Ares-Santos S, Granado N, Espadas I, Martinez-Murillo R, Moratalla R. Methamphetamine causes degeneration of dopamine cell bodies and terminals of the nigrostriatal pathway evidenced by silver staining. Neuropsychopharmacology 2014; 39(5): 1066-80.
[http://dx.doi.org/10.1038/npp.2013.307] [PMID: 24169803]

[41]   Zhu JPQ, Xu W, Angulo JA. Methamphetamine-induced cell death: Selective vulnerability in neuronal subpopulations of the striatum in mice. Neuroscience 2006; 140(2): 607-22.
[http://dx.doi.org/10.1016/j.neuroscience.2006.02.055] [PMID: 16650608]

[42]   Zhu JPQ, Xu W, Angulo JA. Distinct mechanisms mediating methamphetamine-induced neuronal apoptosis and dopamine terminal damage share the neuropeptide substance p in the striatum of mice. Ann N Y Acad Sci 2006; 1074(1): 135-48.
[http://dx.doi.org/10.1196/annals.1369.013] [PMID: 17105911]

[43]   Wang J, Angulo JA. Methamphetamine induces striatal neurokinin-1 receptor endocytosis primarily in somatostatin/NPY/NOS interneurons and the role of dopamine receptors in mice. Synapse 2011; 65(4): 300-8.
[http://dx.doi.org/10.1002/syn.20848] [PMID: 20730802]

[44]   Wang J, Xu W, Ali SF, Angulo JA. Connection between the striatal neurokinin-1 receptor and nitric oxide formation during methamphetamine exposure. Ann N Y Acad Sci 2008; 1139(1): 164-71.
[http://dx.doi.org/10.1196/annals.1432.001] [PMID: 18991860]

[45]   Ramkissoon A, Wells PG. Methamphetamine oxidative stress, neurotoxicity, and functional deficits are modulated by nuclear factor-E2-related factor 2. Free Radic Biol Med 2015; 89: 358-68.
[http://dx.doi.org/10.1016/j.freeradbiomed.2015.07.157] [PMID: 26427884]

[46]   Lee M, Leskova W, Eshaq RS, Harris NR. Acute changes in the retina and central retinal artery with methamphetamine. Exp Eye Res 2020 Apr; 193: 107964. Epub 2020 Feb 7.
[http://dx.doi.org/10.1016/j.exer.2020.107964] [PMID: 32044305] [PMCID: PMC7113125]

[47]   Guo D, Huang X, Xiong T, *et al.* Molecular mechanisms of programmed cell death in methamphetamine induced neuronal damage. Front Pharmacol 2022; 13: 980340.
[http://dx.doi.org/10.3389/fphar.2022.980340] [PMID: 36059947]

[48]   Liu Y, Wen D, Gao J, *et al.* Methamphetamine induces GSDME-dependent cell death in hippocampal neuronal cells through the endoplasmic reticulum stress pathway. Brain Res Bull 2020; 162: 73-83.
[http://dx.doi.org/10.1016/j.brainresbull.2020.06.005] [PMID: 32544512]

[49]   Shah A, Kumar A. Methamphetamine-mediated endoplasmic reticulum (ER) stress induces type-1 programmed cell death in astrocytes *via* ATF6, IRE1α and PERK pathways. Oncotarget 2016; 7(29): 46100-19.
[http://dx.doi.org/10.18632/oncotarget.10025] [PMID: 27323860]

[50]   Zeng Q, Xiong Q, Zhou M, *et al.* Resveratrol attenuates methamphetamine-induced memory impairment *via* inhibition of oxidative stress and apoptosis in mice. J Food Biochem 2021; 45(2): e13622.
[http://dx.doi.org/10.1111/jfbc.13622] [PMID: 33502009]

[51]   Kim B, Yun J, Park B. Methamphetamine induced neuronal damage: Neurotoxicity and neuroinflammation. Biomol Ther 2020; 28(5): 381-8.
[http://dx.doi.org/10.4062/biomolther.2020.044] [PMID: 32668144]

[52]   Garmabi B, Mohaddes R, Rezvani F, Mohseni F, Khastar H, khaksari M. Erythropoietin improve spatial memory impairment following methamphetamine neurotoxicity by inhibition of apoptosis, oxidative stress and neuroinflammation in CA1 area of hippocampus. J Chem Neuroanat 2022; 124: 102137.
[http://dx.doi.org/10.1016/j.jchemneu.2022.102137] [PMID: 35842017]

[53]   Shafahi M, Vaezi G, Shajiee H, Sharafi S, Khaksari M. Crocin inhibits apoptosis and astrogliosis of hippocampus neurons against methamphetamine neurotoxicity *via* antioxidant and anti-inflammatory mechanisms. Neurochem Res 2018; 43(12): 2252-9.
[http://dx.doi.org/10.1007/s11064-018-2644-2] [PMID: 30259275]

[54]   Ghanbari F, Khaksari M, Vaezi G, Hojati V, Shiravi A. Hydrogen sulfide protects hippocampal neurons against methamphetamine neurotoxicity *via* inhibition of apoptosis and neuroinflammation. J Mol Neurosci 2019; 67(1): 133-41.
[http://dx.doi.org/10.1007/s12031-018-1218-8] [PMID: 30456731]

[55]   Che X, Bai Y, Cai J, *et al.* Hippocampal neurogenesis interferes with extinction and reinstatement of methamphetamine associated reward memory in mice. Neuropharmacology 2021; 196: 108717.
[http://dx.doi.org/10.1016/j.neuropharm.2021.108717] [PMID: 34273388]

[56]   Dang J, Tiwari SK, Agrawal K, Hui H, Qin Y, Rana TM. Glial cell diversity and methamphetamine induced neuroinflammation in human cerebral organoids. Mol Psychiatry 2021; 26(4): 1194-207.
[http://dx.doi.org/10.1038/s41380-020-0676-x] [PMID: 32051547]

[57]   Huang J, Ding J, Wang X, *et al.* Transfer of neuron-derived α-synuclein to astrocytes induces neuroinflammation and blood–brain barrier damage after methamphetamine exposure: Involving the regulation of nuclear receptor associated protein 1. Brain Behav Immun 2022; 106: 247-61.
[http://dx.doi.org/10.1016/j.bbi.2022.09.002] [PMID: 36089218]

[58]   Meng Y, Ding J, Li C, Fan H, He Y, Qiu P. Transfer of pathological α-synuclein from neurons to astrocytes *via* exosomes causes inflammatory responses after METH exposure. Toxicol Lett 2020; 331: 188-99.
[http://dx.doi.org/10.1016/j.toxlet.2020.06.016] [PMID: 32569805]

[59]   Ding J, Lian Y, Meng Y, *et al.* The effect of α-synuclein and Tau in methamphetamine induced neurotoxicity *in vivo* and *in vitro*. Toxicol Lett 2020; 319: 213-24.
[http://dx.doi.org/10.1016/j.toxlet.2019.11.028] [PMID: 31783120]

[60]   Meng Y, Qiao H, Ding J, *et al.* Effect of Parkin on methamphetamine-induced α-synuclein degradation dysfunction *in vitro* and *in vivo*. Brain Behav 2020; 10(4): e01574.
[http://dx.doi.org/10.1002/brb3.1574] [PMID: 32086884]

[61]   Ding J, Huang J, Xia B, *et al.* Transfer of α-synuclein from neurons to oligodendrocytes triggers myelin sheath destruction in methamphetamine administration mice. Toxicol Lett 2021; 352: 34-45.
[http://dx.doi.org/10.1016/j.toxlet.2021.09.005] [PMID: 34562559]

[62]   Huang E, Huang H, Guan T, *et al.* Involvement of C/EBPβ-related signaling pathway in

methamphetamine-induced neuronal autophagy and apoptosis. Toxicol Lett 2019; 312: 11-21.
[http://dx.doi.org/10.1016/j.toxlet.2019.05.003] [PMID: 31059759]

[63]   Huang J, Zhang R, Wang S, *et al.* Methamphetamine and HIV-Tat protein synergistically induce oxidative stress and blood-brain barrier damage *via* transient receptor potential melastatin 2 channel. Front Pharmacol 2021; 12: 619436.
[http://dx.doi.org/10.3389/fphar.2021.619436] [PMID: 33815104]

[64]   Northrop NA, Yamamoto BK. Methamphetamine effects on blood-brain barrier structure and function. Front Neurosci 2015; 9: 69.
[http://dx.doi.org/10.3389/fnins.2015.00069] [PMID: 25788874]

[65]   Martins T, Baptista S, Gonçalves J, *et al.* Methamphetamine transiently increases the blood–brain barrier permeability in the hippocampus: Role of tight junction proteins and matrix metalloproteinase-9. Brain Res 2011; 1411: 28-40.
[http://dx.doi.org/10.1016/j.brainres.2011.07.013] [PMID: 21803344]

[66]   Xue Y, He JT, Zhang KK, Chen LJ, Wang Q, Xie XL. Methamphetamine reduces expressions of tight junction proteins, rearranges F-actin cytoskeleton and increases the blood brain barrier permeability *via* the RhoA/ROCK-dependent pathway. Biochem Biophys Res Commun 2019; 509(2): 395-401.
[http://dx.doi.org/10.1016/j.bbrc.2018.12.144] [PMID: 30594393]

[67]   Gonçalves J, Leitão RA, Higuera-Matas A, *et al.* Extended access methamphetamine self administration elicits neuroinflammatory response along with blood brain barrier breakdown. Brain Behav Immun 2017; 62: 306-17.
[http://dx.doi.org/10.1016/j.bbi.2017.02.017] [PMID: 28237710]

[68]   Northrop NA, Yamamoto BK. Persistent neuroinflammatory effects of serial exposure to stress and methamphetamine on the blood brain barrier. J Neuroimmune Pharmacol 2012; 7(4): 951-68.
[http://dx.doi.org/10.1007/s11481-012-9391-y] [PMID: 22833424]

[69]   Northrop NA, Yamamoto BK. Cyclooxygenase activity contributes to the monoaminergic damage caused by serial exposure to stress and methamphetamine. Neuropharmacology 2013; 72: 96-105.
[http://dx.doi.org/10.1016/j.neuropharm.2013.04.040] [PMID: 23643743]

[70]   Tata DA, Raudensky J, Yamamoto BK. Augmentation of methamphetamine-induced toxicity in the rat striatum by unpredictable stress: Contribution of enhanced hyperthermia. Eur J Neurosci 2007; 26(3): 739-48.
[http://dx.doi.org/10.1111/j.1460-9568.2007.05688.x] [PMID: 17686046]

[71]   Eyerman DJ, Yamamoto BK. A rapid oxidation and persistent decrease in the vesicular monoamine transporter 2 after methamphetamine. J Neurochem 2007; 103(3): 1219-27.
[http://dx.doi.org/10.1111/j.1471-4159.2007.04837.x] [PMID: 17683483]

CHAPTER 6

Morphine

Abstract: Morphine is a Schedule II drug and it is used in pain treatment. Like other opioid drugs, it also has addictive properties. The other street names of Morphine include M, Miss Emma, Monkey, Roxanol, and White Stuff. There are natural alkaloids come from the resin of opium poppy, *Papaver somniferum.* Morphine is attached to the receptors in the brain and spinal cord to block pain signals. Morphine impacts the level of dopamine and serotonin by acting in the brain's reward system. Breathing and heart rate are also modified due to morphine which has both short-term and long-term effects. Morphine can last up to 4 to 6 hours in blood. It also acts on the dendrites and spines in order to change the plasticity of the neurons. Endogenous and exogenous opiates target the same tissues and cells.

Keywords: Alkaloids, Dendrites and spines, Endogenous and exogenous opiates, *Papaver somniferum.*

INTRODUCTION

Opium family of compounds includes morphine and other well-known naturally available or synthesized products such as heroin, codeine, hydrocodone, oxycodone, and oxymorphone. Opioids are used as one of the main drugs used in modern medicine. To alleviate severe pain, opioids are the main drugs used. Synthetic opioids such as Fentanyl, meperidine, methadone, and loperamide are also prescribed for medicinal purposes [1]. In both the adult and developing brain, the use of morphine causes changes in the postsynaptic terminals of the excitatory synapses in the limbic system. In adult zebrafish (*Danio rerio*), morphine exposure leads to increased *oprm1* and *npas4a* mRNA levels in the medial zone of the dorsal telencephalon (Dm), ventral region of the ventral telencephalon (Vv), preoptic area [2].

Moreover, morphine leads to poor thinking ability, changes the activity in the brain stem and spinal cord, alters the brain's ability to respond to attacks on the microbes, and harms the memory-making process. Both morphine and fentanyl cause increased intracranial pressure and decreased mean arterial pressure and cerebral profusion pressure [3]. Reference memory and working memory were affected in rats exposed to morphine. Emotional reactivity and anxiety were also reduced in the addicted group in comparison with the normal group [4]. Opioid exposure leads to effects on the amygdala, decreased mu-opioid receptor sensitivity, $GABA_A$ receptor modulation, and modifications in glutamate receptor signalling [5, 6].

Prescription of opioids causes structural and functional changes in reward and motivational areas [7]. In addition, cytoskeletal-related proteins are also affected due to morphine exposure in reward-related areas [8]. There was a difference in the brain metabolic state in regions such as cortex, hypothalamus, brainstem, and cerebellum after morphine consumption in Lewis and Fischer 344 rat strains [9]. It was found that the metabolism of Morphine is much higher in the brain areas of reward in the LEW strain than in the F344 strain. Methadone, an opioid affects early mylenation in the developing rat brain [10]. Drugs of abuse can induce C-FOS expression during withdrawal symptoms, and this C-FOS expression can also be correlated with conditioned place aversion, locomotor sensitization, and conditioned place preference (CPP) [11 - 13]. The FOS expression is noted in the central and basolateral amygdala (CeA and BLA), cingulate cortex, nucleus accumbens (NAc), and bed nucleus of the stria terminalis (BNST) [14].

Prolonged morphine taking caused an upregulation of Bax protein and proapoptotic caspase-3 in rats followed by a reduction in antiapoptotic Bcl-2 protein in rats [15] μ-opioid tolerance has been established by the development of opioid tolerance *via* NMDA receptors [16 - 19]. Synaptic balance is disrupted due to morphine use in the hippocampus [20]. Neural pathways of learning and memory are interconnected with addiction pathways. In patients with cancer and healthy individuals, morphine use interferes with learning and memory [21, 22]. Morphine treatment in chronic conditions caused a reduction (25%) in the perimeter of the ventral tegmental area dopamine neurons [23] (D1-receptor (D1R) expressing and dopamine D2-receptor (D2R) neurons in the nucleus accumbans firing ability was altered due to morphine use [24] In animal models, the injection of morphine at sites, medulla, substantia nigra, nucleus accumbens, and periaqueductal gray (PAG) has caused a reduction in pain behaviour [25].

Under medical supervision, injectable morphine is released into the skin, muscles, and veins. However, some users may misuse morphine as like other opioids without proper medical supervision. It's unavoidable for them as they take this morphine regularly even if it is interfering with their personal and professional work. There can be severe withdrawal symptoms if morphine intake is suddenly stopped. Morphine impacts the brain in various ways such as diminished reflexes, reduced neuroplasticity, impairment in psychomotor function, problems in the amygdala, and disrupted brain synapses causing impaired memory. White matter and cerebellar injury were seen in preterm infants in infants with a birth weight of < 1500 g [26]. In addition, alteration in plasticity was noted in the developing and adult brains in the limbic areas [27].

For opioid disorder methadone (Dolophine) and buprenorphine are the treatment options. Naltrexone (Revia) is another option as it prevents the binding of opioids and blocks their actions. Medication-assisted treatment options can improve survival and reduce opiate misuse. Behavioural therapy includes developing healthy life skills, continuing medications, and changing the attitude of drug misuse. In a model of homocysteine-induced oxidative stress in rat brains, it was shown that morphine increases apoptosis in the hippocampus [28]. Nerve repair and nerve damage in rat models of acute morphine exposure have shown that endoplasmic reticulum stress in the involved [29].

In rat hippocampus and neuroblastoma SH-SY5Y cells, and showed that beclin 1 level are responsible for autophagy [30]. In Morphine tolerance dysfunction of mitophagy is seen and it is due to PINK1/PARKin mediated pathways [31]. Cerebellar brain slices from rats exposed to morphine protected these neurons from ischemia-reperfusion conditions [32]. Microglial activity and neurotoxicity were inhibited by morphine indicating morphine's anti-inflammatory and neuroprotective effects [33]. Morphine tolerance is established by NLTP3 inflammasomes and toll-like receptor 4 [34]. Morphine treatments to microglial cells and neurons have shown that caspase 3 is involved in morphine-induced apoptosis [35]. Another study states that by upregulating microRNA-181-5p morphine induces apoptosis in hippocampal neurons [36]. During pain signalling in lower afferent nociception pathways, it is shown that synapses become sensitive to morphine [37]. In human brain endothelial cells, morphine exposure causes dysregulation of NRF2 pathways and mitochondria dysfunctions [38].

CONCLUSION

Opium family of compounds includes morphine and other well-known naturally available or synthesized products such as heroin, codeine, hydrocodone, oxycodone, and oxymorphone. Opioids are used as one of the main drugs used in modern medicine. To alleviate severe pain, opioids are the main drugs used. Synthetic opioids such as Fentanyl, meperidine, methadone, and loperamide are also prescribed for medicinal purposes [1]. In both the adult and developing brain, the use of morphine causes changes in the postsynaptic terminals of the excitatory synapses in the limbic system. In adult zebrafish (Danio rerio), morphine exposure leads to increased oprm1 and npas4a mRNA levels in the medial zone of the dorsal telencephalon (Dm), ventral region of the ventral telencephalon (Vv), preoptic area. For opioid disorder methadone (Dolophine) and buprenorphine are the treatment options. Naltrexone (Revia) is another option as it prevents the binding of opioids and blocks their actions. In human brain endothelial cells, morphine exposure causes dysregulation of NRF2 pathways and mitochondria dysfunctions.

REFERENCES

[1]   Brownstein MJ. A brief history of opiates, opioid peptides, and opioid receptors. Proc Natl Acad Sci 1993; 90(12): 5391-3.
        [http://dx.doi.org/10.1073/pnas.90.12.5391] [PMID: 8390660]

[2]   Sivalingam M, Ogawa S, Parhar IS. Mapping of morphine-induced *OPRM1* gene expression pattern in the adult zebrafish brain. Front Neuroanat 2020; 14: 5.
        [http://dx.doi.org/10.3389/fnana.2020.00005] [PMID: 32153369]

[3]   de Nadal M, Munar F, Poca MA, Sahuquillo J, Garnacho A, Rosselló J. Cerebral hemodynamic effects of morphine and fentanyl in patients with severe head injury: Absence of correlation to cerebral autoregulation. Anesthesiology 2000; 92(1): 11-9.
        [http://dx.doi.org/10.1097/00000542-200001000-00008] [PMID: 10638893]

[4]   Famitafreshi H, Karimian M, Marefati N. Long-term morphine addiction reduces neurogenesis and memory performance and alters emotional reactivity and anxiety levels in male rats. Open Access Anim Physiol 2015; 7: 129-36.

[5]   Maher CE, Martin TJ, Childers SR. Mechanisms of mu opioid receptor/G-protein desensitization in brain by chronic heroin administration. Life Sci 2005; 77(10): 1140-54.
        [http://dx.doi.org/10.1016/j.lfs.2005.03.004] [PMID: 15890372]

[6]   Zarrindast MR, Ahmadi S, Haeri-Rohani A, Rezayof A, Jafari MR, Jafari-Sabet M. GABAA receptors in the basolateral amygdala are involved in mediating morphine reward. Brain Res 2004; 1006(1): 49-58.
        [http://dx.doi.org/10.1016/j.brainres.2003.12.048] [PMID: 15047023]

[7]   Upadhyay J, Maleki N, Potter J, *et al.* Alterations in brain structure and functional connectivity in prescription opioid-dependent patients. Brain 2010; 133(7): 2098-114.
        [http://dx.doi.org/10.1093/brain/awq138] [PMID: 20558415]

[8]   Yang X, Wen Y, Zhang Y, *et al.* Dynamic changes of cytoskeleton-related proteins within reward-related brain regions in morphine-associated memory. Front Neurosci 2021; 14: 626348.
        [http://dx.doi.org/10.3389/fnins.2020.626348] [PMID: 33584180]

[9]   Soto-Montenegro ML, García-Vázquez V, Lamanna-Rama N, López-Montoya G, Desco M, Ambrosio E. Neuroimaging reveals distinct brain glucose metabolism patterns associated with morphine consumption in Lewis and Fischer 344 rat strains. Sci Rep 2022; 12 (1).

[10]   Vestal-Laborde AA, Eschenroeder AC, Bigbee JW, Robinson SE, Sato-Bigbee C. The opioid system and brain development: Effects of methadone on the oligodendrocyte lineage and the early stages of myelination. Dev Neurosci 2014; 36(5): 409-21.
        [http://dx.doi.org/10.1159/000365074] [PMID: 25138998]

[11]   Hope BT, Simmons DE, Mitchell TB, Kreuter JD, Mattson BJ. Cocaine-induced locomotor activity and Fos expression in nucleus accumbens are sensitized for 6 months after repeated cocaine administration outside the home cage. Eur J Neurosci 2006; 24(3): 867-75.
        [http://dx.doi.org/10.1111/j.1460-9568.2006.04969.x] [PMID: 16930414]

[12]   Pascual MM, Pastor V, Bernabeu RO. Nicotine-conditioned place preference induced CREB phosphorylation and Fos expression in the adult rat brain. Psychopharmacology 2009; 207(1): 57-71.
        [http://dx.doi.org/10.1007/s00213-009-1630-4] [PMID: 19711055]

[13]   Tolliver BK, Sganga MW, Sharp FR. Suppression of c-fos induction in the nucleus accumbens prevents acquisition but not expression of morphine-conditioned place preference. Eur J Neurosci 2000; 12(9): 3399-406.

[http://dx.doi.org/10.1046/j.1460-9568.2000.00214.x] [PMID: 10998122]

[14] Harris GC, Aston-Jones G. Enhanced morphine preference following prolonged abstinence: Association with increased Fos expression in the extended amygdala. Neuropsychopharmacology 2003; 28(2): 292-9.
[http://dx.doi.org/10.1038/sj.npp.1300037] [PMID: 12589382]

[15] Mao J, Sung B, Ji RR, Lim G. Neuronal apoptosis associated with morphine tolerance: Evidence for an opioid-induced neurotoxic mechanism. J Neurosci 2002; 22(17): 7650-61.
[http://dx.doi.org/10.1523/JNEUROSCI.22-17-07650.2002] [PMID: 12196588]

[16] Marek P, Ben-Eliyahu S, Gold M, Liebeskind JC. Excitatory amino acid antagonists (kynurenic acid and MK-801) attenuate the development of morphine tolerance in the rat. Brain Res 1991; 547(1): 81-8.
[http://dx.doi.org/10.1016/0006-8993(91)90576-H] [PMID: 1860074]

[17] Trujillo KA, Akil H. Inhibition of morphine tolerance and dependence by the NMDA receptor antagonist MK-801. Science 1991; 251(4989): 85-7.
[http://dx.doi.org/10.1126/science.1824728] [PMID: 1824728]

[18] Elliott K, Minami N, Kolesnikov YA, Pasternak GW, Inturrisi CE. The NMDA Receptor antagonists, LY274614 and MK-801, and the nitric oxide synthase inhibitor, NG-nitro-L-arginine, attenuate analgesic tolerance to the mu-opioid morphine but not to kappa opioids. Pain 1994; 56(1): 69-75.
[http://dx.doi.org/10.1016/0304-3959(94)90151-1] [PMID: 7512709]

[19] Mao J, Price DD, Mayer DJ. Thermal hyperalgesia in association with the development of morphine tolerance in rats: Roles of excitatory amino acid receptors and protein kinase C. J Neurosci 1994; 14(4): 2301-12.
[http://dx.doi.org/10.1523/JNEUROSCI.14-04-02301.1994] [PMID: 7908958]

[20] Carhart-Harris RL, Kaelen M, Bolstridge M, *et al.* The paradoxical psychological effects of lysergic acid diethylamide (LSD). Psychol Med 2016; 46(7): 1379-90.
[http://dx.doi.org/10.1017/S0033291715002901] [PMID: 26847689]

[21] Lange RA, Hillis LD. Cardiovascular complications of cocaine use. N Engl J Med 2001; 345(5): 351-8.
[http://dx.doi.org/10.1056/NEJM200108023450507] [PMID: 11484693]

[22] Kurita GP, Sjøgren P, Ekholm O, *et al.* Prevalence and predictors of cognitive dysfunction in opioid-treated patients with cancer: A multinational study. J Clin Oncol 2011; 29(10): 1297-303.
[http://dx.doi.org/10.1200/JCO.2010.32.6884] [PMID: 21357785]

[23] Tavron LS, Wei-xing SHI, Sarah BL, Herbert WH, Benjamin SB, Eric J. NESTLER.Chronic morphine induces visible changes in the morphology of mesolimbic dopamine neurons. Proc Natl Acad Sci 2023; 93: 11202-7.

[24] McDevitt DS, Jonik B, Graziane NM. Morphine differentially alters the synaptic and intrinsic properties of D1R- and D2R-expressing medium spiny neurons in the nucleus accumbens. Front Synaptic Neurosci 2019; 11: 35.
[http://dx.doi.org/10.3389/fnsyn.2019.00035] [PMID: 31920618]

[25] Yaksh TL. Pharmacology and mechanisms of opioid analgesic activity. Acta Anaesthesiol Scand 1997; 41(1): 94-111.
[http://dx.doi.org/10.1111/j.1399-6576.1997.tb04623.x] [PMID: 9061092]

[26] Al-Mouqdad MM, Jamjoom DZ, Huseynova R, *et al.* Association between morphine exposure and impaired brain development on term-equivalent age brain magnetic resonance imaging in very preterm infants. Sci Rep 2022; 12(1): 4498.
[http://dx.doi.org/10.1038/s41598-022-08677-0] [PMID: 35296792]

[27] Beltrán-Campos V, Silva-Vera M, García-Campos ML, Díaz-Cintra S. Effects of morphine on brain plasticity. Neurologia 2015; 30(3): 176-80.

[PMID: 25444409]

[28] Karkhah A, Ataee R, Ataie A. Morphine pre- and post-conditioning exacerbates apoptosis in rat hippocampus cells in a model of homocysteine-induced oxidative stress. Biomed Rep 2017; 7(4): 309-13.
[http://dx.doi.org/10.3892/br.2017.962] [PMID: 28928969]

[29] Liu J, Yi S, Shi W, *et al.* The pathology of morphine-inhibited nerve repair and morphine-induced nerve damage is mediated *via* endoplasmic reticulum stress. Front Neurosci 2021; 15: 618190.
[http://dx.doi.org/10.3389/fnins.2021.618190] [PMID: 33679302]

[30] Zhao L, Zhu Y, Wang D, *et al.* Morphine induces Beclin 1- and ATG5-dependent autophagy in human neuroblastoma SH-SY5Y cells and in the rat hippocampus. Autophagy 2010; 6(3): 386-94.
[http://dx.doi.org/10.4161/auto.6.3.11289] [PMID: 20190558]

[31] Kong H, Jiang CY, Hu L, *et al.* Morphine induces dysfunction of PINK1/Parkin-mediated mitophagy in spinal cord neurons implying involvement in antinociceptive tolerance. J Mol Cell Biol 2019; 11(12): 1056-68.
[http://dx.doi.org/10.1093/jmcb/mjz002] [PMID: 30698724]

[32] Lim YJ, Zheng S, Zuo Z. Morphine preconditions Purkinje cells against cell death under *in vitro* simulated ischemia-reperfusion conditions. Anesthesiology 2004; 100(3): 562-8.
[http://dx.doi.org/10.1097/00000542-200403000-00015] [PMID: 15108969]

[33] Qian L, Tan KS, Wei SJ, *et al.* Microglia-mediated neurotoxicity is inhibited by morphine through an opioid receptor-independent reduction of NADPH oxidase activity. J Immunol 2007; 179(2): 1198-209.
[http://dx.doi.org/10.4049/jimmunol.179.2.1198] [PMID: 17617613]

[34] Qu J, Tao XY, Teng P, *et al.* Blocking ATP-sensitive potassium channel alleviates morphine tolerance by inhibiting HSP70-TLR4-NLRP3-mediated neuroinflammation. J Neuroinflammation 2017; 14(1): 228.
[http://dx.doi.org/10.1186/s12974-017-0997-0] [PMID: 29178967]

[35] Hu S, Sheng WS, Lokensgard JR, Peterson PK. Morphine induces apoptosis of human microglia and neurons. Neuropharmacology 2002; 42(6): 829-36.
[http://dx.doi.org/10.1016/S0028-3908(02)00030-8] [PMID: 12015209]

[36] Wang YL, An XH, Zhang XQ, Liu JH, Wang JW, Yang ZY. Morphine induces the apoptosis of mouse hippocampal neurons HT-22 through upregulating miR-181-5p. Eur Rev Med Pharmacol Sci 2020; 24(12): 7114-21.
[PMID: 32633406]

[37] Pollard KJ, Bowser DA, Anderson WA, Meselhe M, Moore MJ. Morphine-sensitive synaptic transmission emerges in embryonic rat microphysiological model of lower afferent nociceptive signaling. Sci Adv 2021; 7(35): eabj2899.
[http://dx.doi.org/10.1126/sciadv.abj2899] [PMID: 34452921]

[38] Reymond S, Vujić T, Schvartz D, Sanchez JC. Morphine-induced modulation of Nrf2-antioxidant response element signaling pathway in primary human brain microvascular endothelial cells. Sci Rep 2022; 12(1): 4588.
[http://dx.doi.org/10.1038/s41598-022-08712-0] [PMID: 35301408]

CHAPTER 7

Ketamine

Abstract: Ketamine is a dissociative anaesthetic drug that functions as a blocker of NMDA receptors. Moreover, it causes a neurostimulatory effect and is also used as a sedative. Ketamine has many names, such as Special K, Green K, Super K, Super Acid, Jet, and Cat Valium. Ketamine is used as a recreational drug in clubs, also known as a "club drug". As a recreational drug, it causes the patient to experience delirium and an altered state of consciousness. Patients with cardiovascular disabilities can also be given ketamine as a sedative. Ketamine can be taken in various methods, such as orally, rectally, intranasally, IV, IM, or intrathecally. Ketamine abuse can lead to secondary renal damage and upper gastrointestinal symptoms.

Keywords: Anaesthetic drugs, Blockers of NMDA receptors, Sedative recreational drugs.

INTRODUCTION

Ketamine is used with other medications or alone as a general anaesthetic agent. It is a cyclohexane derivative and dissociative drug. It binds to NMDA receptors as a non-competitive antagonist. It can also bind to other receptors that are producing anaesthetic effects, ex.opioid, adrenergic, cholinergic, and monoamine receptors [1]. Ketamine is safely used as an anaesthetic agent worldwide in operation theaters for children and adults [2]. In chronic pain management, ketamine is used [3]. In the liver, ketamine is metabolised by N-demethylation and ring hydroxylation pathways [4]. At Parke Davis Laboratories, in 1963, ketamine was developed as a replacement for phencycline as its chemical nature and mechanism are similar to ketamine. Ketamine does not interact with GABA receptors. Norketamine is the major metabolite of ketamine for metabolism in the liver. The half-life of ketamine is 186 min. Ketamine is used in many forms, such as smoking, mixed in drinks, injection, or liquids that can be mixed in liquids. The side effects of ketamine are depression, flashbacks, hallucinations, agitation, and unconsciousness. Tolerance to ketamine can be established by regular usage. Overdose can cause coma and death. Ketamine bladder syndrome can be established in long-term users along with headaches and stomach pain. Regular using of this drug can lead to problems with memory and concentration. Cardiac arrhythmias can happen in people who overdose on ketamine with alcohol or heroin or with other opioids. Memory loss is reported to be the long-term effect of taking ketamine. In the primary somatosensory cortex, ketamine has inhibited the excitatory synaptic transmission of neurons [5]. (HCN1 pacemaker channels were found to be the substrate for the anesthetic effect of ketamine [6]). In another study, it was found that excitability was impaired by ketamine by blocking sodium and voltage-gated potassium currents in superficial dorsal horn neurons [7].

Ketamine as Antidepressant

Ketamine is a N-methyl-D-aspartate receptor (NMDAR) antagonist, and its effect as an antidepressant has been shown in various research articles. On the other hand, by activating the adult-borne immature granule neurons, ketamine alleviated the depression-like symptoms in mice [8]. Yet in another study, it was interesting to note that ketamine at sub-anaesthetic doses has been shown to treat depression in the clinical trials conducted [9]. Oral rapamycin treatment in patients with depression along with ketamine treatment has raised the level of antidepressant-like activity of ketamine. It means rapamycin has been shown to extend the benefits of ketamine [10]. Eukaryotic elongation factor 2 (eEF2) kinase has been deactivated by ketamine resulting in reduced EF2 phosphorylation causing the antidepressant-like property of ketamine [11]. In human patients, treatment with a single ketamine dose has created a rapid and robust antidepressant response [12].

Ketamine as an Anti-inflammatory Compound

The effect of ketamine on neuroinflammatory changes is because of the given dose. Spatial memory recognition was disturbed by ketamine at a dose of 60mg/kg also in mice, it was shown to reduce anxiety-like behaviour [13]. Single administration and long-term administration of ketamine had tremendous effects on the levels of inflammatory markers such as TNFa in mice. A single dose has induced TNFa levels high, but repeated dose has not increased its levels [13]. Locomotor activity was compared between male and female rats. It was found that 10 mg/kg ketamine has been shown to reduce spontaneous locomotor activity in male rats, whereas 40 mg/kg induced locomotor activity in female rats [14]. Ketamine (20 mg/kg) single dose has not prevented neuro inflammation in rats [15]. Treatment-resistant depression (TRD) is a challenge to the medical world. Ketamine has evidently shown to induce anti-inflammatory actions by causing antidepressant activity [16]. After repeated ketamine exposure, TNFa levels were downregulated in patients [5]. LPS-induced delirium neuro inflammation was attenuated by ketamine. Interestingly, in 2019, Esketamine, a nasal spray was approved by the Federal Drug Administration (FDA) to treat depression [17]. There are studies that found that in comparison with other anti-depressants, ketamine plays a major role in treating depression [18]. Transient receptor potential vanilloid 4 (TRPV4) is known to be involved in various neurodegenerative diseases. C57BL/6 mice were injected with ketamine and proved that ketamine has improved cognitive dysfunction in Perioperative Neurocognitive Disorder (PND) [19]. Ketamine also reduced inflammation in the primary glial cells upon infection with LPS [20]. In rats, it was found that ketamin's effects were seen in reducing depression and this was mediated by the downregulation of proinflammatory cytokines in the rat hippocampus following microglia deactivation [6]

Ketamine Induces Apoptosis in Neuronal Cells

In the rhesus monkey, ketamine has induced cell death *via* the necrotic and apoptotic processes as analysed by electron microscopy. Nuclear condensation and fragmentation were observed in neuronal cells in the rhesus monkey brain [21]. Ketamine at a dose of 1, 10, and 20 µM treatment in hippocampal neurons has shown downregulation of multiple cellular markers [22]. The research also says that an imbalance in calcium signalling pathways in neuronal cells is the main reason for ketamine-induced abnormalities [23]. Moreover, *in vitro* studies have shown that ketamine at a dose of 0, 20, 100, or 500 µM treated for 6 and 24 h in human iPSC-derived dopaminergic neurons disrupted the mitochondrial electron transport system causing autophagy and mitochondrial dysfunction [24] immature γ-aminobutyric acidergic (GABAergic) interneurons were cultured and treated with 5 µg/ml ketamine for 4 hrs. This resulted in alterations in the dendritic growth and the dendritic arbour was impaired. When neural stem cells were treated with ketamine, it caused a significant decrease in viability and reduced proliferation with an increase in apoptosis [25]. Surprisingly, some studies on hippocampal cell lines upon treatment with ketamine have shown an increase in cellular proliferation [26]. On thalamocortical slices, ketamine, it was shown by the patch clamp technique that ketamine suppressed the activity of voltage-gated sodium channels [27].

Adult-born immature granule neurons (ABINs) from the mouse hippocampus when treated with ketamine have shown full activation thus alleviating the depressive behaviour [8]. There are studies that point out that although ketamine causes neural stem cell proliferation, at the same time, it can induce neuronal apoptosis *via* a mechanism involving mitochondrial reactive oxygen species production [28]. In the mouse cerebral cortex, also an apoptosis mechanism was observed after ketamine treatment [29]. In the developing brain in the neural progenitor cells, ketamine alters the neurogenesis process, this finding can be used to educate pregnant women who misuse this drug [30]. Ketamine was found to induce apoptosis in human lymphocytes and neuronal cells at millimolar concentrations, whereas at higher concentrations it was found to induce necrosis. These mechanisms were mediated *via* the mitochondrial pathways [31]. Adult male Wistar rats were treated with 10mg/kg of ketamine. Hypertrophic astrocytes were seen in the ketamine-treated group, followed by neurodegeneration in the CA4 region of the hippocampus [32].

CONCLUSION

Ketamine is an N-methyl-D-aspartate receptor (NMDAR) antagonist, and its effect as an antidepressant has been shown in various research articles. On the other hand, by activating the adult-borne immature granule neurons, ketamine alleviated the depression-like symptoms in mice [8]. Yet in another study, it was interesting to note that ketamine at sub-anaesthetic doses has been shown to treat depression in the clinical trials conducted. LPS-induced delirium neuro inflammation was attenuated by ketamine. Interestingly, in 2019, Esketamine, a nasal spray was approved by the Federal Drug Administration (FDA) to treat depression. In the rhesus monkey, ketamine has induced cell death via the necrotic and apoptotic processes as analysed by electron microscopy. Nuclear condensation and fragmentation were observed in neuronal cells in the rhesus monkey brain. Adult-born immature granule neurons (ABINs) from the mouse hippocampus when treated with ketamine have shown full activation thus alleviating the depressive behaviour.

REFERENCES

[1]   Persson J. Wherefore ketamine? Curr Opin Anaesthesiol 2010; 23(4): 455-60.
       [http://dx.doi.org/10.1097/ACO.0b013e32833b49b3] [PMID: 20531172]

[2]   Green SM, Roback MG, Kennedy RM, Krauss B. Clinical practice guideline for emergency department ketamine dissociative sedation: 2011 update. Ann Emerg Med 2011; 57(5): 449-61.
       [http://dx.doi.org/10.1016/j.annemergmed.2010.11.030] [PMID: 21256625]

[3]   Hocking G, Cousins MJ. Ketamine in chronic pain management: An evidence-based review. Anesth Analg 2003; 97(6): 1730-9.
       [http://dx.doi.org/10.1213/01.ANE.0000086618.28845.9B] [PMID: 14633551]

[4]   Reves JG, Glass PS, Lubarsky DA, McEvoy MD, Ruiz RM. Intravenous anaesthetics. Miller's Anaesthesia. 7th ed.. Miller RD. USA: Churchill Livingstone 2010; p. 719. Available from: https://pubmed.ncbi.nlm.nih.gov/?term=5.%09Reves+JG%2C+Glass+PS%2C+Lubarsky+DA%2C+McEvoy+MD%2C+Ruiz+RM.+Intravenous+anaesthetics.+In%3A+Miller+RD%2C+editor.+Miller%27s+Anaesthesia.+7th+ed.+USA%3A+Churchill+Livingstone%3B+2010.+pp.+719%E2%80%9371.+

[5]   Zhan Y, Zhou Y, Zheng W, *et al*. Alterations of multiple peripheral inflammatory cytokine levels after repeated ketamine infusions in major depressive disorder. Transl Psychiatry 2020; 10(1): 246.
       [http://dx.doi.org/10.1038/s41398-020-00933-z] [PMID: 32699226]

[6]   Chen X, Shu S, Bayliss DA. HCN1 channel subunits are a molecular substrate for hypnotic actions of ketamine. J Neurosci 2009; 29(3): 600-9.
       [http://dx.doi.org/10.1523/JNEUROSCI.3481-08.2009] [PMID: 19158287]

[7]   Schnoebel R, Wolff M, Peters SC, *et al*. Ketamine impairs excitability in superficial dorsal horn neurones by blocking sodium and voltage-gated potassium currents. Br J Pharmacol 2005; 146(6):

826-33.
[http://dx.doi.org/10.1038/sj.bjp.0706385] [PMID: 16151436]

[8]   Rawat R, Tunc-Ozcan E, McGuire TL, Peng CY, Kessler JA. Ketamine activates adult-born immature granule neurons to rapidly alleviate depression-like behaviors in mice. Nat Commun 2022; 13(1): 2650.
[http://dx.doi.org/10.1038/s41467-022-30386-5] [PMID: 35551462]

[9]   Berman RM, Cappiello A, Anand A, *et al.* Antidepressant effects of ketamine in depressed patients. Biol Psychiatry 2000; 47(4): 351-4.
[http://dx.doi.org/10.1016/S0006-3223(99)00230-9] [PMID: 10686270]

[10]   Abdallah CG, Averill LA, Gueorguieva R, *et al.* Modulation of the antidepressant effects of ketamine by the mTORC1 inhibitor rapamycin. Neuropsychopharmacology 2020; 45(6): 990-7.
[http://dx.doi.org/10.1038/s41386-020-0644-9] [PMID: 32092760]

[11]   Autry AE, Adachi M, Nosyreva E, *et al.* NMDA receptor blockade at rest triggers rapid behavioural antidepressant responses. Nature 2011; 475(7354): 91-5.
[http://dx.doi.org/10.1038/nature10130] [PMID: 21677641]

[12]   Zarate CA Jr, Brutsche NE, Ibrahim L, *et al.* Replication of ketamine's antidepressant efficacy in bipolar depression: a randomized controlled add-on trial. Biol Psychiatry 2012; 71(11): 939-46.
[http://dx.doi.org/10.1016/j.biopsych.2011.12.010] [PMID: 22297150]

[13]   Li Y, Shen R, Wen G, *et al.* Effects of ketamine on levels of inflammatory cytokines IL-6, IL-1β, and TNF-α in the hippocampus of mice following acute or chronic administration. Front Pharmacol 2017; 8 (MAR).
[http://dx.doi.org/10.3389/fphar.2017.00139]

[14]   Spencer HF, Berman RY, Boese M, *et al.* Effects of an intravenous ketamine infusion on inflammatory cytokine levels in male and female Sprague–Dawley rats. J Neuroinflammation 2022; 19(1): 75.
[http://dx.doi.org/10.1186/s12974-022-02434-w] [PMID: 35379262]

[15]   Moraga-Amaro R, Guerrin CGJ, Reali Nazario L, *et al.* A single dose of ketamine cannot prevent protracted stress-induced anhedonia and neuroinflammation in rats. Stress 2022; 25(1): 145-55.
[http://dx.doi.org/10.1080/10253890.2022.2045269] [PMID: 35384793]

[16]   Nikkheslat N. Targeting inflammation in depression: Ketamine as an anti-inflammatory antidepressant in psychiatric emergency. Brain Behav Immun Health 2021; 18: 100383.

[17]   Choudhury D, Autry AE, Tolias KF, Krishnan V. Ketamine: Neuroprotective or Neurotoxic? Front Neurosci 2021; 15: 672526.
[http://dx.doi.org/10.3389/fnins.2021.672526] [PMID: 34566558]

[18]   Zanos P, Gould TD. Mechanisms of ketamine action as an antidepressant. Mol Psychiatry 2018; 23(4): 801-11.
[http://dx.doi.org/10.1038/mp.2017.255] [PMID: 29532791]

[19]   Li Q, Zhou DN, Tu YQ, Wu XW, Pei DQ, Xiong Y. Ketamine administration ameliorates anesthesia and surgery-induced cognitive dysfunction *via* activation of TRPV4 channel opening. Exp Ther Med 2022; 24(1): 478.
[http://dx.doi.org/10.3892/etm.2022.11405] [PMID: 35761804]

[20]   Shibakawa YS, Sasaki Y, Goshima Y, *et al.* Effects of ketamine and propofol on inflammatory responses of primary glial cell cultures stimulated with lipopolysaccharide. Br J Anaesth 2005; 95(6): 803-10.
[http://dx.doi.org/10.1093/bja/aei256] [PMID: 16227338]

[21]   Slikker W Jr, Zou X, Hotchkiss CE, *et al.* Ketamine-induced neuronal cell death in the perinatal rhesus monkey. Toxicol Sci 2007; 98(1): 145-58.
[http://dx.doi.org/10.1093/toxsci/kfm084] [PMID: 17426105]

[22]   Pichl T, Keller T, Hünseler C, *et al.* Effects of ketamine on neurogenesis, extracellular matrix homeostasis and proliferation in hypoxia-exposed HT22 murine hippocampal neurons. Biomed Rep 2020; 13(4): 23.

[23]   Lisek M, Zylinska L, Boczek T. Ketamine and calcium signaling-a crosstalk for neuronal physiology and pathology. Int J Mol Sci 2020; 21(21): 8410.
[http://dx.doi.org/10.3390/ijms21218410] [PMID: 33182497]

[24]   Ito H, Uchida T, Makita K. Ketamine causes mitochondrial dysfunction in human induced pluripotent stem cell-derived neurons. PLoS One 2015; 10(5): e0128445.
[http://dx.doi.org/10.1371/journal.pone.0128445] [PMID: 26020236]

[25]   Lu P, Lei S, Li W, *et al.* Dexmedetomidine protects neural stem cells from ketamine-induced injury. Cell Physiol Biochem 2018; 47(4): 1377-88.
[http://dx.doi.org/10.1159/000490823] [PMID: 29929189]

[26]   Peters AJ, Villasana LE, Schnell E. Ketamine alters hippocampal cell proliferation and improves learning in mice after traumatic brain injury. Anesthesiology 2018; 129(2): 278-95.
[http://dx.doi.org/10.1097/ALN.0000000000002197] [PMID: 29734230]

[27]   Yin J, Fu B, Wang Y, Yu T. Effects of ketamine on voltage-gated sodium channels in the barrel cortex and the ventral posteromedial nucleus slices of rats. Neuroreport 2019; 30(17): 1197-204.
[http://dx.doi.org/10.1097/WNR.0000000000001344] [PMID: 31568204]

[28]   Bai X, Yan Y, Canfield S, *et al.* Ketamine enhances human neural stem cell proliferation and induces neuronal apoptosis *via* reactive oxygen species-mediated mitochondrial pathway. Anesth Analg 2013; 116(4): 869-80.
[http://dx.doi.org/10.1213/ANE.0b013e3182860fc9] [PMID: 23460563]

[29]   Wang Q, Shen F, Zou R, Zheng J, Yu X, Wang Y. Ketamine-induced apoptosis in the mouse cerebral cortex follows similar characteristic of physiological apoptosis and can be regulated by neuronal activity. Mol Brain 2017; 10(1): 24.
[http://dx.doi.org/10.1186/s13041-017-0302-2] [PMID: 28623920]

[30]   Dong C, Anand KJS. Developmental neurotoxicity of ketamine in pediatric clinical use. Toxicol Lett 2013; 220(1): 53-60.
[http://dx.doi.org/10.1016/j.toxlet.2013.03.030] [PMID: 23566897]

[31]   Braun S, Gaza N, Werdehausen R, *et al.* Ketamine induces apoptosis *via* the mitochondrial pathway in human lymphocytes and neuronal cells. Br J Anaesth 2010; 105(3): 347-54.
[http://dx.doi.org/10.1093/bja/aeq169] [PMID: 20659914]

[32]   Yuan C, Zhang Y, Zhang Y. Effects of ketamine on neuronal spontaneous excitatory postsynaptic currents and miniature excitatory postsynaptic currents in the somatosensory cortex of rats. Iran J Med Sci 41; (4): 275-82.

CHAPTER 8

Fentanyl

Abstract: Fentanyl is an opioid usually used in general anaesthesia, due to which it is also called an analgesic drug. These drugs can relieve the pain within the body by blocking the neurotransmitters or chemicals that cause pain in the body. Opioids can work in both the ascending pathways of the brain as well as the descending pathways of the brain for pain modulation. Fentanyl is more potent than morphine and herion. Fentanyl is also given as transdermal patches or lozenges in the treatment of pain management. Fentanyl is also sold illegally and can cause of death too when abused. Because of its strong property to be addicted, fentanyl also is mixed with the heroine. Moreover, fentanyl has its own effects during withdrawal, which causes behaviour changes. Fentanyl can bind to μ-opioid receptors (MORs) to exert its effects. In addition, fentanyl abuse is becoming more common globally. Fentanyl causes the brain to suffocate by decreasing the oxygen supply, causing hypoxia and hyperglycemia as well. Fentanyl abuse can cause serious cognitive issues, leading to severe structural damage manifested as hormonal and neuronal disturbances. By suppressing the two brainstem areas, opioids cause disturbances to breathing.

Keywords: Addiction, Decreasing oxygen supply, Hyperglycemia, Opioid, μ-opioid receptors (MORs).

INTRODUCTION

Long-Evans rats, when exposed to fentanyl, showed decreased oxygen supply in the basolateral amygdale as revealed by electrochemical detection [1]. At 3–7 μg/kg dose range, fentanyl causes significant hypoxic effects in human patients [2]. In the dorsolateral pons, due to fentanyl, the respiratory neurons can be affected so breathing is stopped [3]. Fentanyl also causes muscular rigidity by acting on neurons in the area known as locus ceruleus. This area of the brain is responsible for regulating adrenaline levels in the brain. Brain malondialdehyde levels were enhanced by fentanyl treatment [4]. In preterm infants, fentanyl administration protects the developing brain by relieving pain [5].

Fentanyl in the powdered form looks quite similar to other drugs. Fentanyl can be mixed with cocaine, heroin, and methamphetamines. Fentanyl is always mixed with heroin, cocaine, and methamphetamine and sold as pills to look very similar to other prescription opioids. Lacing fentanyl with other drugs is often lethal but the users are unaware of this.

Hippocampus and Fentanyl

In the Schaffer-collateral CA1 pathway in the hippocampus, fentanyl treatment disturbs long-term potentiation and enhances long-term depression [6]. Apart from this, fentanyl at lower concentrations, such as 0.01 and 0.1 μM, can decrease AMPA receptors and cause the destruction of dendritic spines in cultured hippocampal neurons. As the hippocampus plays an essential role in memory, this effect can cause problems in memory storage [7]. Not only that opioids when administered chronically caused the impairment of cognitive function [8], but cortical neurons, and glial cells, were co-cultured and treated with low (0.01 μM) and high (10 μM) doses of fentanyl caused gene expression changes in the synaptic plasticity property [9]. Moreover, this kind of exposure to fentanyl also damages somatosensory circuit behaviour and functions [10]. During early development periods, in the 2nd and 3rd layer of somatosensory neurons, male and female C57BL/6J mice, when given environmental enrichment, have shown changes in function [11].

Neurotoxic Effect of Fentanyl

Some studies point out that not only the adrenaline level but also the glutamate neuronal network is involved in the muscular rigidity caused by fentanyl [12]. By using extracellular recordings, the rat brain slice preparation has been used to understand the effect of serotonin and noradrenaline (NA), on nociception during fentanyl and morphine treatment. The firing activities of these opioids are inhibited by fentanyl and morphine treatment as well [13]. Fentanyl was found to be neurotoxic to spinal cord neurons [14]. Fentanyl also acts on the GABAergic neurons in the vagus nerve to increase parasympathetic neurotransmission by decreasing GABAergic transmission [15]. This kind of inhibition was seen in both the pre- and postsynaptic neurons.

Noradrenaline and Fentanyl Receptors

In two types of cells, such as human neuroblastoma SHSY5Y and rat phaeochromocytoma PC12, exposure to fentanyl caused a decrease in noradrenaline uptake [16]. There are studies that fentanyl not only acts as amu–opioid receptor but it also acts as an agonist on the 5-HT1A receptor. Fentanyl overdose can also be due to this effect of binding to both receptors [17]. There are a number of evidence that opioids cause disturbances to adult neurogenesis in the dentate gyrus region of the hippocampus, leading to the cessation of differentiation and maturation process in the hippocampus [18].

Mitochondria Damage and Fentanyl

Neuroblastoma/glioma hybrid cell line was investigated for the effect of fentanyl, methadone, and morphine if they cause changes in mitochondrial function. The mitochondrial network was decreased by fentanyl and methadone treatment but not by morphine [19]. In glioma cells, there are reverse findings that morphine causes changes in the mitochondrial membrane potential but not methadone and fentanyl [20]. In the nucleus accumbens (NAc) there are changes in the mitochondrial copy number that are reflected in blood leukocytes as well [21]. In the human neuroblastoma cell line SH-SY5Y, it was demonstrated that fentanyl causes cell death by both necrosis and apoptosis mechanisms [22]. When human hepatoma HepG2 cells were exposed to fentanyl, it was found that mitochondria were greatly affected, which means the respiration rate of mitochondria was affected [23].

Fentanyl and Pain-related and Responsive Areas –PET Studies

Positron emission tomography (PET) studies have shown that fentanyl in human subjects has reduced the blood supply to the thalamus and posterior cingulate cortex [24]. Fentanyl significantly and selectively affects the cerebral brain areas that are associated with pain-related and responsive areas [25]. Mid-anterior cingulate cortex area is activated during fentanyl administration as revealed by PET imaging in human subjects. So this part of the brain may be involved in the fentanyl-mediated analgesic effects [26]. Pain-related areas and pathways in the cingulate cortex and orbitofrontal cortex have shown an increase in the regional blood supply during fentanyl administration in human subjects. These areas are responsible for addiction, nociception, and reward behaviours. Some of the studies point out that Fentanyl administration increased cerebral blood perfusion (CBP) in brain areas where mu-opioid receptors are present [27]. This study is further supported by another study that remifentanil administration has increased the blood supply to areas where mu-opioid receptors are present. In contrast, the primary somatosensory cortex has shown less activation as it contains lower mu-opioid receptors [28]. Fentanyl withdrawal is dangerous as it increases anxiety-like behaviour [29]. Male rats prefer fentanyl than food choice, indicating even in humans, men are more addictive to fentanyl than women [30]. Fentanyl causes respiratory depression, in turn leading to hypoxia and hyperglycemia leading to changes in brain metabolism and temperature in rats [1]. In rats exposed to fentanyl, electroencephalogram, and auditory evoked potentials were changed during fentanyl treatment [31]. There are studies that prove that heroin contaminated with fentanyl causes brain hypothermia and brain hypoxia. Fentanyl-induced seizures cause changes in the brain and cerebral blood flow in rat brains. Nucleus accumbent mediates the behavioural changes associated with fentanyl induced changes in the brain [32]. Effect of fentanyl on hippocampal neurons were dose-dependent. At low concentrations, it decreased AMPA receptors and decreased the number of dendritic spines. At high concentrations, it induces new spines and increases the number of AMPA receptors [7]. *In vitro*, in rat vagus nerve C fibers at isotonic concentrations, fentanyl blocked the nerve conduction [33]. EEG has revealed fentanyl stops people breathing and becomes dangerous to their lives. In freely moving rats, fentanyl administration caused decreased glutamate and increased GABA levels [34]. There are differences in antinociceptive signaling in rats exposed to morphine and fentanyl. Both have caused signal ign through mu-opioid receptors but for morphine blocking of RGS proteins causes more signaling through mu-opioid receptors [35]. Under deep brain stimulation technique in the subthalamic nucleus it was found that fentanyl can be used for advanced Parkinson's disease treatment [36, 37]. In the cortico-striatal-thalamo-cortical (CSTC) circuit, fentanyl causes changes in the circuitary by leading to addiction and behaviour changes in human primates [37]. Microinjection of SP600125, JNKinhibitor, into the periaqeductal grey area has induced the development of tolerance to fentanyl in rats [38]. Fentanyl increased the P-glycoprotein (P-gp) membrane ATPase activity in the blood-brain barrier [39]. Fentanyl showed strong nociceptive action during cortical activity [40]. In rats, cerebral blood flow and metabolic rate for oxygen and seizures were studied after fentanyl administration. It was found that fentanyl has increased the metabolic rate for oxygen during seizures [41]. Affective and somatic withdrawal signs and brain reward function reduction are due to fentanyl withdrawal in rats [42]. Negative signs of fentanyl administration depend on the dose of fentanyl but not on the duration of fentanyl administration [43]. Rats injected with fentanyl were neurotoxic to spinal cord neurons [14] (Abut *et al.*, 2015).

CONCLUSION

Fentanyl in the powdered form looks quite similar to other drugs. Fentanyl can be mixed with cocaine, heroin, and methamphetamines. Fentanyl is always mixed with heroin, cocaine, and methamphetamine and sold as pills to look very similar to other prescription opioids. Lacing fentanyl with other drugs is often lethal but the users are unaware of this. Some studies point out that not only the adrenaline level but also the glutamate neuronal network is involved in the muscular rigidity caused by fentanyl. By using extracellular recordings, the rat brain slice preparation has been used to understand the effect of serotonin and noradrenaline (NA), on nociception during fentanyl and morphine treatment. The firing activities of these opioids are inhibited by fentanyl and morphine treatment as well. Neuroblastoma/glioma hybrid cell line was investigated for the effect of fentanyl, methadone, and morphine if they cause changes in mitochondrial function. The mitochondrial network was decreased by fentanyl and methadone treatment but not by morphine.

REFERENCES

[1] Solis E Jr, Cameron-Burr KT, Kiyatkin EA. Heroin Contaminated with Fentanyl Dramatically Enhances Brain Hypoxia and Induces Brain Hypothermia. eNeuro 2017; 4(5): ENEURO.0323-17.2017.
[http://dx.doi.org/10.1523/ENEURO.0323-17.2017] [PMID: 29085909]

[2] Dahan A, Yassen A, Bijl H, *et al.* Comparison of the respiratory effects of intravenous buprenorphine and fentanyl in humans and rats. Br J Anaesth 2005; 94(6): 825-34.
[http://dx.doi.org/10.1093/bja/aei145] [PMID: 15833777]

[3] Saunders SE, Baekey DM, Levitt ES. Fentanyl effects on respiratory neuron activity in the dorsolateral pons. J Neurophysiol 2022; 128(5): 1117-32.
[http://dx.doi.org/10.1152/jn.00113.2022] [PMID: 36197016]

[4] Yadav SK, Nagar DP, Bhattacharya R. Effect of fentanyl and its three novel analogues on biochemical, oxidative, histological, and neuroadaptive markers after sub-acute exposure in mice. Life Sci 2020; 246: 117400.
[http://dx.doi.org/10.1016/j.lfs.2020.117400] [PMID: 32032645]

[5] Qiu J, Zhao L, Yang Y, Zhang J, Feng Y, Cheng R. Effects of fentanyl for pain control and neuroprotection in very preterm newborns on mechanical ventilation. J Matern Fetal Neonatal Med 2019; 32(22): 3734-40.
[http://dx.doi.org/10.1080/14767058.2018.1471593] [PMID: 29712500]

[6] Tian H, Xu Y, Liu F, Wang G, Hu S. Effect of acute fentanyl treatment on synaptic plasticity in the hippocampal CA1 region in rats. Front Pharmacol 2015; 6(OCT): 251.
[http://dx.doi.org/10.3389/fphar.2015.00251] [PMID: 26578961]

[7] Lin H, Higgins P, Loh HH, Law PY, Liao D. Bidirectional effects of fentanyl on dendritic spines and AMPA receptors depend upon the internalization of mu opioid receptors. Neuropsychopharmacology 2009; 34(9): 2097-111.
[http://dx.doi.org/10.1038/npp.2009.34] [PMID: 19295508]

[8] Ersche KD, Clark L, London M, Robbins TW, Sahakian BJ. Profile of executive and memory function associated with amphetamine and opiate dependence. Neuropsychopharmacology 2006; 31(5): 1036-47.
[http://dx.doi.org/10.1038/sj.npp.1300889] [PMID: 16160707]

[9] Lam D, Sebastian A, Bogguri C, *et al.* Dose-dependent consequences of sub-chronic fentanyl exposure on neuron and glial co-cultures. Front Toxicol 2022; 4: 983415.
[http://dx.doi.org/10.3389/ftox.2022.983415]

[10] Alipio JB, Haga C, Fox ME, *et al.* Perinatal fentanyl exposure leads to long-lasting impairments in somatosensory circuit function and behavior. J Neurosci 2021; 41(15): 3400-17.
[http://dx.doi.org/10.1523/JNEUROSCI.2470-20.2020] [PMID: 33853934]

[11] Alipio JB, Riggs LM, Plank M, Keller A. Environmental enrichment mitigates the long-lasting sequelae of perinatal fentanyl exposure in mice. J Neurosci 2022; 42(17): 3557-69.
[http://dx.doi.org/10.1523/JNEUROSCI.2083-21.2022] [PMID: 35332082]

[12] Fu MJ, Tsen LY, Lee TY, Lui PW, Chan SHH. Involvement of cerulospinal glutamatergic neurotransmission in fentanyl-induced muscular rigidity in the rat. Anesthesiology 1997; 87(6): 1450-9.
[http://dx.doi.org/10.1097/00000542-199712000-00024] [PMID: 9416730]

[13] Alojado MES, Ohta Y, Yamamura T, Kemmotsu O. The effect of fentanyl and morphine on neurons in the dorsal raphe nucleus in the rat: An *in vitro* study. Anesth Analg 1994; 78(4): 726-32.

[http://dx.doi.org/10.1213/00000539-199404000-00019]

[14]   Abut YC, Turkmen AZ, Midi A, Eren B, Yener N, Nurten A. Neurotoxic effects of levobupivacaine and fentanyl on rat spinal cord. Rev Bras Anestesiol 2015; 65(1): 27-33.
[http://dx.doi.org/10.1016/j.bjan.2013.07.005] [PMID: 25497746]

[15]   Griffioen KJS, Venkatesan P, Huang ZG, *et al.* Fentanyl inhibits GABAergic neurotransmission to cardiac vagal neurons in the nucleus ambiguus. Brain Res 2004; 1007(1-2): 109-15.
[http://dx.doi.org/10.1016/j.brainres.2004.02.010] [PMID: 15064141]

[16]   Atcheson R, Rowbotham DJ, Lambert DG. Fentanyl inhibits the uptake of [3H]noradrenaline in cultured neuronal cells. Br J Anaesth 1993; 71(4): 540-3.
[http://dx.doi.org/10.1093/bja/71.4.540] [PMID: 8260304]

[17]   Tao R, Karnik M, Ma Z, Auerbach SB. Effect of fentanyl on 5-HT efflux involves both opioid and 5-HT$_{1A}$ receptors. Br J Pharmacol 2003; 139(8): 1498-504.
[http://dx.doi.org/10.1038/sj.bjp.0705378] [PMID: 12922937]

[18]   Zhang Y, Loh HH, Law PY. Effect of opioid on adult hippocampal neurogenesis. ScientificWorldJournal 2016; 2016: 2601264.
[http://dx.doi.org/10.1155/2016/2601264]

[19]   Nylander E, Zelleroth S, Nyberg F, Grönbladh A, Hallberg M. The effects of morphine, methadone, and fentanyl on mitochondria: A live cell imaging study. Brain Res Bull 2021; 171: 126-34.
[http://dx.doi.org/10.1016/j.brainresbull.2021.03.009] [PMID: 33741459]

[20]   Mastronicola D, Arcuri E, Arese M, *et al.* Morphine but not fentanyl and methadone affects mitochondrial membrane potential by inducing nitric oxide release in glioma cells. Cell Mol Life Sci 2004; 61(23): 2991-7.
[http://dx.doi.org/10.1007/s00018-004-4371-x] [PMID: 15583861]

[21]   Calarco CA, Fox ME, Van Terheyden S, *et al.* Mitochondria-related nuclear gene expression in the nucleus accumbens and blood mitochondrial copy number after developmental fentanyl exposure in adolescent male and female C57BL/6 mice. Front Psychiatry 2021; 12: 737389.
[http://dx.doi.org/10.3389/fpsyt.2021.737389] [PMID: 34867530]

[22]   Sogos V, Caria P, Porcedda C, *et al.* Human neuronal cell lines as an *in vitro* toxicological tool for the evaluation of novel psychoactive substances. Int J Mol Sci 2021; 22(13): 6785.
[http://dx.doi.org/10.3390/ijms22136785] [PMID: 34202634]

[23]   Djafarzadeh S, Vuda M, Jeger V, Takala J, Jakob SM. The effects of fentanyl on hepatic mitochondrial function. Anesth Analg 2016; 123(2): 311-25.
[http://dx.doi.org/10.1213/ANE.0000000000001280] [PMID: 27089001]

[24]   Adler LJ, Gyulai FE, Diehl DJ, Mintun MA, Winter PM, Firestone LL. Regional brain activity changes associated with fentanyl analgesia elucidated by positron emission tomography. Anesth Analg 1997; 84(1): 120-6.
[http://dx.doi.org/10.1213/00000539-199701000-00023] [PMID: 8989012]

[25]   Firestone LL, Gyulai F, Mintun M, Adler LJ, Urso K, Winter PM. Human brain activity response to fentanyl imaged by positron emission tomography. Anesth Analg 1996; 82(6): 1247-51.
[PMID: 8638799]

[26]   Casey KL, Svensson P, Morrow TJ, Raz J, Jone C, Minoshima S. Selective opiate modulation of nociceptive processing in the human brain. J Neurophysiol 2000; 84(1): 525-33.
[http://dx.doi.org/10.1152/jn.2000.84.1.525] [PMID: 10899224]

[27]   Zelaya FO, Zois E, Muller-Pollard C, *et al.* The response to rapid infusion of fentanyl in the human brain measured using pulsed arterial spin labelling. MAGMA 2012; 25(2): 163-75.
[http://dx.doi.org/10.1007/s10334-011-0293-4] [PMID: 22113518]

[28]   Leppä M, Korvenoja A, Carlson S, *et al.* Acute opioid effects on human brain as revealed by functional magnetic resonance imaging. Neuroimage 2006; 31(2): 661-9.

[http://dx.doi.org/10.1016/j.neuroimage.2005.12.019] [PMID: 16459107]

[29]   Fujii K, Koshidaka Y, Adachi M, Takao K. Effects of chronic fentanyl administration on behavioral characteristics of mice. Neuropsychopharmacol Rep 2019; 39(1): 17-35.
[http://dx.doi.org/10.1002/npr2.12040] [PMID: 30506634]

[30]   Townsend EA, Negus SS, Caine SB, Thomsen M, Banks ML. Sex differences in opioid reinforcement under a fentanyl *vs*. food choice procedure in rats. Neuropsychopharmacology 2019; 44(12): 2022-9.
[http://dx.doi.org/10.1038/s41386-019-0356-1] [PMID: 30818323]

[31]   Antunes LM, Roughan JV, Flecknell PA. Excitatory effects of fentanyl upon the rat electroencephalogram and auditory-evoked potential responses during anaesthesia. Eur J Anaesthesiol 2003; 20(10): 800-8.
[http://dx.doi.org/10.1097/00003643-200310000-00005] [PMID: 14580049]

[32]   Long JJ, Ma J, Stan Leung L. Behavioral depression induced by an amygdala seizure and the opioid fentanyl was mediated through the nucleus accumbens. Epilepsia 2009; 50(8): 1953-61.
[http://dx.doi.org/10.1111/j.1528-1167.2009.02143.x] [PMID: 19490038]

[33]   Power I, Brown DT, Wildsmith JA. The effect of fentanyl, meperidine and diamorphine on nerve conduction *in vitro*. Reg Anesth 1991; 16(4): 204-8.
[PMID: 1911495]

[34]   Pourzitaki C, Tsaousi G, Papazisis G, *et al*. Fentanyl and naloxone effects on glutamate and GABA release rates from anterior hypothalamus in freely moving rats. Eur J Pharmacol 2018; 834: 169-75.
[http://dx.doi.org/10.1016/j.ejphar.2018.07.029] [PMID: 30030987]

[35]   Morgan MM, Tran A, Wescom RL, Bobeck EN. Differences in antinociceptive signalling mechanisms following morphine and fentanyl microinjections into the rat periaqueductal gray. Eur J Pain 2020; 24(3): 617-24.
[http://dx.doi.org/10.1002/ejp.1513] [PMID: 31785128]

[36]   Kim W, Song IH, Lim YH, *et al*. Influence of propofol and fentanyl on deep brain stimulation of the subthalamic nucleus. J Korean Med Sci 2014; 29(9): 1278-86.
[http://dx.doi.org/10.3346/jkms.2014.29.9.1278] [PMID: 25246748]

[37]   Withey SL, Cao L, de Moura FB, *et al*. Fentanyl-induced changes in brain activity in awake nonhuman primates at 9.4 Tesla. Brain Imaging Behav 2022; 16(4): 1684-94.
[http://dx.doi.org/10.1007/s11682-022-00639-4] [PMID: 35226333]

[38]   Morgan MM, Reid RA, Saville KA. Functionally selective signaling for morphine and fentanyl antinociception and tolerance mediated by the rat periaqueductal gray. PLoS One 2014; 9(12): e114269.
[http://dx.doi.org/10.1371/journal.pone.0114269] [PMID: 25503060]

[39]   Yu C, Yuan M, Yang H, Zhuang X, Li H. P-glycoprotein on blood-brain barrier plays a vital role in fentanyl brain exposure and respiratory toxicity in rats. Toxicol Sci 2018; 164(1): 353-62.
[http://dx.doi.org/10.1093/toxsci/kfy093] [PMID: 29669042]

[40]   Peng YZ, Li XX, Wang YW. Effects of Parecoxib and Fentanyl on nociception-induced cortical activity. Mol Pain 2010; 6: 1744-8069-6-3.
[http://dx.doi.org/10.1186/1744-8069-6-3] [PMID: 20089200]

[41]   Carlsson C, Smith DS, Keykhah MM, Englebach I, Harp JR. The effects of high-dose fentanyl on cerebral circulation and metabolism in rats. Anesthesiology 1982; 57(5): 375-80.
[http://dx.doi.org/10.1097/00000542-198211000-00005] [PMID: 7137618]

[42]   Bruijnzeel AW, Lewis B, Bajpai LK, Morey TE, Dennis DM, Gold M. Severe deficit in brain reward function associated with fentanyl withdrawal in rats. Biol Psychiatry 2006; 59(5): 477-80.
[http://dx.doi.org/10.1016/j.biopsych.2005.07.020] [PMID: 16169528]

[43]   Liu J, Pan H, Gold MS, Derendorf H, Bruijnzeel AW. Effects of fentanyl dose and exposure duration on the affective and somatic signs of fentanyl withdrawal in rats. Neuropharmacology 2008; 55(5): 812-8.
[http://dx.doi.org/10.1016/j.neuropharm.2008.06.034] [PMID: 18634811]

CHAPTER 9

Alcohol

Abstract: Alcohol affects brain activity in various ways. It has both short-term and long-term effects. It causes slurred speech, short-term memory dysfunctions hallucinations, *etc.* by timing the activity of neuronal cells. Moreover, it causes teratogenic effects in the fetus ifthe mother is consuming alcohol during pregnancy. Alcohol can damage the brain cells, cause a lowering of serotonin levels, and higher GABA levels, cease new brain cells to be formed, and cause damage to the blood vessels and nerve cells in the brain. In addition, alcohol abuse causes Wernicke-Korsakoff's syndrome, which is due to the lack of vitamin B1 in drinkers. Also, alcohol abuse causes Wernicke's encephalopathy which is characterised by muscle problems, being confused, *etc.* Memory loss and less coordination are the long-term effects of alcohol abuse. All regions of the brain, such as the cerebellum, limbic system, and cerebral cortex, can be affected by alcohol abuse. The cerebellum is responsible for the movement of the body, and alcohol disrupts this balance causing emotional and memory issues. Alcohol consumption on a regular basis leads to reduced brain size or a rapid aging process. Alcohol disorder is listed as one of the most prevalent mental health problems in the world.

Keywords: Alcohol disorder, Memory loss and less coordination, Serotonin levels, Teratogenic effects in the fetus.

INTRODUCTION

A study conducted in Russia states that vodka use is the major risk factor for death for various reasons in young adults [1]. Mental disorders, cognitive dysfunctions, dementia, and learning and memory defects are common in alcohol users. Alcohol is a depressant; it should be withheld from children and people under the age of 18. Alcohol is very harmful to the developing brain. It can cause the developing brain to be affected in such a way that problem-solving skills will be affected. Seizures, stroke, and dementia are common problems due to alcohol usage. Alcohol also causes damage to the dendrites, which are branch-like structures arising from brain cells. The brain scans of alcohol abusers have shown a high reduction of grey matter volume. Excessive alcohol consumption also causes neurodegeneration. Alcohol use disorder (AUD) is a disorder caused by excessive alcohol abuse causing emotional disturbances. AUD has vast effects on

Jayalakshmi Krishnan
All rights reserved-© 2024 Bentham Science Publishers

neuronal cells in terms of disturbed synaptic contact, blood-brain barrier dysfunction, demyelination, dementia, *etc.* [2]. Dementia is characterised by psychological changes caused by chronic alcohol use [3]. Acetaldehyde is a metabolite of alcohol, and when the brain cells are exposed to this metabolite, a decrease in growth factors is observed that are leading to neuronal death or degeneration [4]. Alcohol suppresses the communication between nerve cells. Alcohol increases chloride ion conductance inside the cells, making it more negative. Because it binds with the GABA receptor, it opens the chloride ion channels to allow more negative charge inside the cell which the cell becomes a hyperpolarised state. An individual who depends on alcohol can also develop a tolerance to alcohol as it causes intoxication. Alcohol abuse also causes a person to lose social life, loss of jobs, loss of interpersonal skills, liver failure, neurotoxicity, *etc.* [5]. By AUD, 3.3 million annual deaths have been reported by WHO due to alcohol neurotoxicity. In human postmortem brain samples, at the dentate gyrus and subgranular zone, it was shown that persons who died from going alcohol abuse have reduced the number of neurogenic pools *i.e.*, stem cell pools [6]. Abstinence from alcohol causes the brain to revert to its neurogenesis process. Prolonged alcohol intake disrupts the excitatory neurotransmitter pathways [7]. Cytokines and proinflammatory mediators are released in response to binge ethanol treatment under TLR4 activation in adolescent mice.

Role of Toll-like Receptors in Ethanol-induced Changes

There are studies in which alcohol is known to activate microglia and astrocytes through toll-like receptors leading to neuroinflammation [8]. The behavioural and cognitive dysfunction associated with alcohol is due to the epigenetic changes caused by TLR4 activation [9]. There are evidences that scute exposure to ethanol in TLR4++ mice causes microgial cell activation in the brain in comparison with TLR4(-/-) mice. Ethanol induces the expression of TLR4 and TLR2 in microglial cells causing inflammation [10]. Another mechanism by which ethanol activates the TLR is due to TLR4/IL-1RI responses leading to inflammation [11]. Mice defect in TLR4 is protected from various kinds of molecular and behavioural changes, thus further confirming that TLR plays a role in inflammation [12]. In line with the same observations, it was found that TLR 4 knockout mice were prevented from autophagy during ethanol treatment [13]. Ethanol-induced brain damage is due to the recruitment of TLR4/IL-1RI to membrane lipid rafts, causing inflammatory activation [14]. Signaling of these receptors is known to cause cell death in astrocytes treated with ethanol [15]. These receptors are endocytosed in lipid rafts in astroglial cells [16]. Clathrin-dependent pathways or lipid raft caveolae are involved in this type of endocytosis mechanism [9]. When adolescent rats are exposed to high doses of ethanol neuroimmune changes and myelin changes do happen [17]. In the rat hippocampus, SUMO-specific protease

6 (SENP6) and TLR4 are involved in ethanol-induced neuroinflammation [18]. TLR4 activation causes impairment of ubiquitin pathways in mice treated with ethanol in the cerebral cortex [19]. Astrocytes-derived extracellular vesicles are involved in TLR4-mediated neuroinflammation in cortical neurons and astrocyte cultures [20]. In alcohol-fed WT and TLR4(-/-) mice, It was demonstrated that Blood Brain barrier was damaged due to the infiltration of leukocytes [21]. This is mediated by TLR/NLRP3 neuroinflammatory responses.

Ethanol and Animal Studies

In order to under the effect of ethanol on adolescent brains, male rats were treated with different doses of ethanol such as 1.0, 2.5, or 5.0 g/kg, i.g. The brains of these rats have shown in the dentate gyrus there was a reduced neurogenesis [22]. Sprague-Dawley rats (8-12 g/kg/d) were exposed to ethanol for 4 days. The perfused brain samples revealed impaired neurogenesis and brain damage in the cortico-limbic region [23]. In developing rat brains, it was found (GD 17.5) that ethanol causes apoptotic neurodegeneration in hippocampal and primary cortical neurons [24]. Cytotoxic edema was also found to be associated with chronic ethanol exposure in rat brains [25]. Alcohol introduction after abstinence in rat models causes oxidative stress and it induces neuroinflammation [26]. Tyrosine hydroxylase is an enzyme responsible for dopamine synthesis. The levels of this enzyme were higher in the ventral tegmental area of Sprague-Dawley rats. This finding can be useful to understand the rewarding properties of ethanol in the rat brain [27]. Adolescent male Wistar rats exposed to ethanol have shown motor impairments in the open field, behavioural tasks, and cerebral cortex damage [28]. Neurodegeneration and cognitive defects were noted in preclinical rat experiments [22]. Neuronal loss is also noted in the prefrontal cortex of the brain due to alcohol exposure. The evidence suggest that α-Synuclein is involved in the structural changes associated with alcohol exposure in the human brain [29]. Brain atrophy is also another effect of alcohol relapse in patients, especially in areas with behavioural control [30]. There are studies that found that neurotransmitter diffusion is also hampered in the brain due to chronic alcohol consumption [31]. Ethanol causes physical dependence in rats after abstinence for days [32]. This is due to the effect of ethanol on brain ribosomes causing altered protein synthesis [33]. Further studies state that mRNA binding sites in ribosomes are affected due to ethanol intake which leads to tolerance [34]. Mice developing tolerance to ethanol, effects were seen in the orbitofrontal cortex (OFC) of the neurons [35]. Altered long-term potentiation in the medial prefrontal cortex of mice exposed to ethanol was also noted [36]. Toll-like receptor-mediated neuroinflammation is also involved in brain injury following ethanol exposure [21]. In the anterior lobe of the cerebellum, there was a reduction in Purkinje cells in rats when exposed to ethanol [37]. Gliosis induced by ethanol in the rat brain can also be used as a marker for alcohol exposure and its effects [38]. Binge ethanol exposure leads to decreased neurogenesis and corticolimbic region degeneration in Adult Sprague–Dawley rats [39]. This can lead to the understanding of neuroinflammatory events related to binge ethanol exposure. Neuronal apoptosis was observed in neonatal rats exposed to ethanol and chlorogenic acid exposure prevents this kind of ethanol-induced apoptosis [40]. Neuronal degeneration in the prefrontal cortex was observed in mice upon exposure to subchronic alcohol. Ethanol-induced neurodegeneration is seen in rats very commonly [41]. In adult male rats, binge ethanol exposure causes neuroinflammation and neurodegeneration in the hippocampus, entorhinal cortex, and olfactory bulb [42].

Effects in Humans

Alcohol is a CNS depressant. It causes a lot of problems in humans when consumed. For example, memory problems, speech problems, loss of control, loss of coordination, vision problems, *etc.* Severe alcohol consumption can lead to problems in breathing, irregular heartbeat, stop, of normal heart function, seizures, low body temperature, dehydration, *etc.* Alcohol neuropathy can also be noted during the excessive drinking of alcohol. This is manifested as bladder problems, urination problems, cramps, and abnormal changes in bowel function. Severe alcohol consumption leads to Wernicke encephalopathy[v], which is due to Vitamin B1 deficiency. This condition can also result in mental dysfunction and coma.

Korsakoff psychosis is the sequential event of vitamin B1 deficiency. Auditory or visual hallucinations can also be caused due to this syndrome in alcoholic patients. Heavy alcohol use leads to reduced white matter areas, reduced synapses, increased neuroinflammation, *etc.* [43]. RNA splicing was altered due to alcohol use in the human brain [44]. Human embryonic stem cell (hESC)-cortical neurons showed upregulated NMDA receptor genes when exposed to ethanol, leading to a suggestion of alcohol-induced neuroadaptation [45]. In the peripheral blood samples, alcohol-induced alteration in the genes involved in apoptosis, proliferation, and DNA repair was observed. These observations can be used to understand the changes in the behavioural pattern of humans exposed to alcohol use disorder [46]. In a study conducted to understand the brain metabolites after ethanol, by using magnetic resonance (MR) spectroscopic analysis, it was found that creatine, choline, inositol, and aspartate levels were decreased [47]. In postmortem brains of alcoholics, excessive neuroinflammation was observed and the same finding was observed in animal brains [48]. In adult rat brains phospholipase enzyme A2 levels were decreased due to ethanol exposure, in organotypic slice cultures phospholipases A2 were decreased followed by oxidative stress, neuroinflammation, and neurodegeneration due to binge ethanol exposure [42, 49, 50].

Ethanol and Alzheimer's Disease Model

3xTg-AD mouse model displayed proinflammation due to binge ethanol exposure in the adolescent stage, paving the way to early Alzheimer's disease pathology [51]. Also, chronic exposure to binge alcohol causes lysosomal impairment in female 3xTg-AD mice and causes Tu protein pathology in the cortical and hippocampal brain regions [52]. Using 3xTg-AD, mice were exposed to alcohol and it was observed that they developed AD-like symptoms in the brain. There was a reduction in Akt/mTOR signalling pathways in the entorhinal cortex and hippocampus as well [53]. In adolescent rats exposed to ethanol, there was a loss in hippocampal neurogenesis and it was blocked by anti-inflammatory drugs [54]. Impaired cognitive function was due to the loss of neurogenesis in the hippocampus in rats that are exposed to binge ethanol [55]. The same observation is seen in Wistar rats exposed to ethanol leading to the loss of adult hippocampal neurogenesis [56]. Apart from these, there are neuroimmune and epigenetic changes also involved in ethanol-induced inflammatory changes in rat brains [57]. There is a decrease in histone H3 acetylation in the hippocampus as an effect of adolescent alcohol exposure [58]. In Amygdale, adolescent intermittent ethanol modulates enzymes involved in H3K9 dimethylation responsible for chromatin remodelling and pschopathology [59]. Dysregulation of DNA methylation in the amygdala is seen due to alcohol exposure in the adolescent periods which even persists up to adulthood [60]. Binge alcohol drinking during adolescence leads to abnormal brain development during adulthood [61]. Adult hippocampal neurogenesis is impaired due to secretion of proinflammatory molecules in microglial cells due to ethanol exposure [62]. In adult male Wistar rats, through CB1 cannabinoid receptor ethanol inhibits adult hippocampal neurogenesis [63]. When both adolescent and adult rats were exposed to ethanol, it was found that increased cell death and loss of hippocampal neurogenesis were observed in adolescent rats [64]. Wistar rats when exposed to intermittent ethanol exposure during adolescence time, there was a decrease in hippocampal neurogenesis [65]. In male Wistar rats, it was found that alcohol exposure during adulthood causes changes in neuropeptide Y pathways *via* histone acetylation in the brain [66]. Cholinergic and neuroimmune signaling is involved in the alcohol-induced pathology of adult neurogenesis [67 - 73]. There is a loss of cholinergic forebrain neurons due to adolescent binge ethanol exposure [54].

CONCLUSION

Mental disorders, cognitive dysfunctions, dementia, and learning and memory defects are common in alcohol users. Alcohol is a depressant; it should be withheld from children and people under the age of 18. Alcohol is very harmful to the developing brain. It can cause the developing brain to be affected in such a way that problem-solving skills will be affected. Seizures, stroke, and dementia are common problems due to alcohol usage. Alcohol also causes damage to the dendrites, which are branch-like structures arising from brain cells. The brain scans of alcohol abusers have shown a high reduction of grey matter volume. In order to under the effect of ethanol on adolescent brains, male rats were treated with different doses of ethanol such as 1.0, 2.5, or 5.0 g/kg, i.g. The brains of these rats have shown in the dentate gyrus there was a reduced neurogenesis [22]. Sprague-Dawley rats (8-12 g/kg/d) were exposed to ethanol for 4 days. The perfused brain samples revealed impaired neurogenesis and brain damage in the cortico-limbic region.

REFERENCES

[1]   Zaridze D, Lewington S, Boroda A, *et al.* Alcohol and mortality in Russia: Prospective observational study of 151 000 adults. Lancet 2014; 383(9927): 1465-73.
        [http://dx.doi.org/10.1016/S0140-6736(13)62247-3] [PMID: 24486187]

[2]   Lieber CS, Victor M. The effects of alcohol on the nervous system. Med Nutr Complicat Alcohol 1992; pp. 413-57.

[3]   Kwok CL. Central nervous system neurotoxicity of chronic alcohol abuse. Asia Pac J Med Toxicol 2016; 2: 70-1.

[4]   Tabrizi S. Neurodegenerative diseases neurobiology pathogenesis and therapeutics. J Neurol Neurosurg Psychiatry 2006; 77(2): 284.
        [http://dx.doi.org/10.1136/jnnp.2005.072710]

[5]   Bose J, Hedden SL, Lipari RN, Park-Lee E, Porter J, Pemberton M. Key substance use and mental health indicators in the United States: Results from the 2015 national survey on drug use and health. report of the substance abuse and mental health services administration. 2016.

[6]   Le Maître TW, Dhanabalan G, Bogdanovic N, Alkass K, Druid H. Effects of alcohol abuse on proliferating cells, stem/progenitor cells, and immature neurons in the adult human hippocampus. Neuropsychopharmacology 2018; 43(4): 690-9.
        [http://dx.doi.org/10.1038/npp.2017.251] [PMID: 29052615]

[7]   Crews FT, Mdzinarishvili A, Kim D, He J, Nixon K. Neurogenesis in adolescent brain is potently inhibited by ethanol. Neuroscience 2006; 137(2): 437-45.
        [http://dx.doi.org/10.1016/j.neuroscience.2005.08.090] [PMID: 16289890]

[8]   Crews F, Nixon K, Kim D, *et al.* BHT blocks NF-kappaB activation and ethanol-induced brain damage. Alcohol Clin Exp Res 2006; 30(11): 1938-49.
        [http://dx.doi.org/10.1111/j.1530-0277.2006.00239.x] [PMID: 17067360]

[9]   Naseer MI, Ullah I, Narasimhan ML, *et al.* Neuroprotective effect of osmotin against ethanol-induced apoptotic neurodegeneration in the developing rat brain. Cell Death Dis 2014; 5(3): e1150.
        [http://dx.doi.org/10.1038/cddis.2014.53] [PMID: 24675468]

[10]   Liu H, Zheng W, Yan G, *et al.* Acute ethanol-induced changes in edema and metabolite concentrations

in rat brain. BioMed Res Int 2014; 2014: 1-8.
[http://dx.doi.org/10.1155/2014/351903] [PMID: 24783201]

[11] Fernández-Rodríguez S, Cano-Cebrián MJ, Rius-Pérez S, *et al.* Different brain oxidative and neuroinflammation status in rats during prolonged abstinence depending on their ethanol relapse-like drinking behavior: Effects of ethanol reintroduction. Drug Alcohol Depend 2022; 232: 109284.
[http://dx.doi.org/10.1016/j.drugalcdep.2022.109284] [PMID: 35033958]

[12] Lee YK, Park SW, Kim YK, *et al.* Effects of naltrexone on the ethanol-induced changes in the rat central dopaminergic system. Alcohol Alcohol 2005; 40(4): 297-301.
[http://dx.doi.org/10.1093/alcalc/agh163] [PMID: 15897221]

[13] Teixeira FB, Santana LNS, Bezerra FR, *et al.* Chronic ethanol exposure during adolescence in rats induces motor impairments and cerebral cortex damage associated with oxidative stress. PLoS One 2014; 9(6): e101074.
[http://dx.doi.org/10.1371/journal.pone.0101074] [PMID: 24967633]

[14] Crews FT, Nixon K. Mechanisms of neurodegeneration and regeneration in alcoholism. Alcohol Alcohol 2009; 44(2): 115-27.
[http://dx.doi.org/10.1093/alcalc/agn079] [PMID: 18940959]

[15] Janeczek J, Lewohi . Chapter 8 - Effect of alcohol on the regulation of α-synuclein in the human brain. Addictive Substances and Neurological Disease Alcohol, Tobacco, Caffeine, and Drugs of Abuse in Everyday Lifestyles . 2017; pp. 67-73.

[16] Beck A, Wu T Stenberg, Genauck A, *et al.* Effect of brain structure, brain function, and brain connectivity on relapse in alcohol-dependent patients Arch Gen Psychiatry 2012; 69 (8).

[17] De Santis S, Cosa-Linan A, Garcia-Hernandez R, *et al.* Chronic alcohol consumption alters extracellular space geometry and transmitter diffusion in the brain. Sci Adv 2020; 6(26): eaba0154.
[http://dx.doi.org/10.1126/sciadv.aba0154] [PMID: 32637601]

[18] Tewari S, Sweeney FM, Fleming EW. Ethanol-induced changes in properties of rat brain ribosomes. Neurochem Res 1980; 5(9): 1025-35.
[http://dx.doi.org/10.1007/BF00966140] [PMID: 7193807]

[19] Tewari S, Goldstein MA, Noble EP. Alterations in cell free brain protein synthesis following ethanol withdrawal in physically dependent rats. Brain Res 1977; 126(3): 509-18.
[http://dx.doi.org/10.1016/0006-8993(77)90601-1] [PMID: 558814]

[20] Tewari S, Greenberg SA, Do K, Grey PA. The response of rat brain protein synthesis to ethanol and sodium barbital. Alcohol Drug Res 1987; 7(4): 243-58.
[PMID: 3828001]

[21] De la Monte SM, Kril JJ. Human alcohol-related neuropathology. Acta Neuropathol 2014; 127(1): 71-90.
[http://dx.doi.org/10.1007/s00401-013-1233-3] [PMID: 24370929]

[22] Nimitvilai S, Lopez MF, Mulholland PJ, Woodward JJ. Chronic intermittent ethanol exposure enhances the excitability and synaptic plasticity of lateral orbitofrontal cortex neurons and induces a tolerance to the acute inhibitory actions of ethanol. Neuropsychopharmacology 2016; 41(4): 1112-27.
[http://dx.doi.org/10.1038/npp.2015.250] [PMID: 26286839]

[23] Kroener S, Mulholland PJ, New NN, Gass JT, Becker HC, Chandler LJ. Chronic alcohol exposure alters behavioral and synaptic plasticity of the rodent prefrontal cortex. PLoS One 2012; 7(5): e37541.
[http://dx.doi.org/10.1371/journal.pone.0037541] [PMID: 22666364]

[24] Van Booven D, Mengying Li , Sunil Rao J, *et al.* Alcohol use disorder causes global changes in splicing in the human brain. Transl Psychiatry 2021; 11(1): 2.
[http://dx.doi.org/10.1038/s41398-020-01163-z] [PMID: 33414398]

[25] Xiang Y, Kim KY, Gelernter J, Park IH, Zhang H. Ethanol upregulates NMDA receptor subunit gene expression in human embryonic stem cell-derived cortical neurons. PLoS One 2015; 10(8): e0134907.

[http://dx.doi.org/10.1371/journal.pone.0134907] [PMID: 26266540]

[26] Hicks SD, Lewis L, Ritchie J, *et al.* Evaluation of cell proliferation, apoptosis, and dna-repair genes as potential biomarkers for ethanol-induced cns alterations. BMC Neurosci 2012; 13(1): 128.
[http://dx.doi.org/10.1186/1471-2202-13-128] [PMID: 23095216]

[27] Biller A, Bartsch AJ, Homola G, Solymosi L, Bendszus M. The effect of ethanol on human brain metabolites longitudinally characterized by proton MR spectroscopy. J Cereb Blood Flow Metab 2009; 29(5): 891-902.
[http://dx.doi.org/10.1038/jcbfm.2009.12] [PMID: 19240741]

[28] Alfonso-Loeches S, Pascual-Lucas M, Blanco AM, Sanchez-Vera I, Guerri C. Pivotal role of TLR4 receptors in alcohol-induced neuroinflammation and brain damage. J Neurosci 2010; 30(24): 8285-95.
[http://dx.doi.org/10.1523/JNEUROSCI.0976-10.2010] [PMID: 20554880]

[29] Ghosh B, Sharma R, Yadav S, Parashar V, Jagdish P. Ethanol exposure induces cerebellar neuronal loss in rats. Eur J Anat 2020; 24(5): 407-13.

[30] Hayes DM, Deeny MA, Shaner CA, Nixon K. Determining the threshold for alcohol-induced brain damage: New evidence with gliosis markers. Alcohol Clin Exp Res 2013; 37(3): 425-34.
[http://dx.doi.org/10.1111/j.1530-0277.2012.01955.x] [PMID: 23347220]

[31] Crews F, Nixon K, Kim D, *et al.* BHT blocks NF-kappaB activation and ethanol-induced brain damage. Alcohol Clin Exp Res 2006; 30(11): 1938-49.
[http://dx.doi.org/10.1111/j.1530-0277.2006.00239.x] [PMID: 17067360]

[32] Guo Z, Li J. Chlorogenic acid prevents alcohol-induced brain damage in neonatal rat. Transl Neurosci 2017; 8 (1).
[http://dx.doi.org/10.1515/tnsci-2017-0024]

[33] Saito M, Chakraborty G, Hui M, Masiello K, Saito M. Ethanol-induced neurodegeneration and glial activation in the developing brain. Brain Sci 2016; 6(3): 31.
[http://dx.doi.org/10.3390/brainsci6030031] [PMID: 27537918]

[34] Mukherjee S. Alcoholism and its effects on the central nervous system. Curr Neurovasc Res 2013; 10(3): 256-62.
[http://dx.doi.org/10.2174/15672026113109990004] [PMID: 23713737]

[35] Montesinos J, Alfonso-Loeches S, Guerri C. Impact of the innate immune response in the actions of ethanol on the central nervous system. Alcohol Clin Exp Res 2016; 40(11): 2260-70.
[http://dx.doi.org/10.1111/acer.13208] [PMID: 27650785]

[36] Yang JY, Xue X, Tian H, *et al.* Role of microglia in ethanol-induced neurodegenerative disease: Pathological and behavioral dysfunction at different developmental stages. Pharmacol Ther 2014; 144(3): 321-37.
[http://dx.doi.org/10.1016/j.pharmthera.2014.07.002] [PMID: 25017304]

[37] Pascual M, Baliño P, Alfonso-Loeches S, Aragón CMG, Guerri C. Impact of TLR4 on behavioral and cognitive dysfunctions associated with alcohol-induced neuroinflammatory damage. Brain Behav Immun 2011; 25 (Suppl. 1): S80-91.
[http://dx.doi.org/10.1016/j.bbi.2011.02.012] [PMID: 21352907]

[38] Fernandez-Lizarbe S, Montesinos J, Guerri C. Ethanol induces TLR 4/ TLR 2 association, triggering an inflammatory response in microglial cells. J Neurochem 2013; 126(2): 261-73.
[http://dx.doi.org/10.1111/jnc.12276] [PMID: 23600947]

[39] Fernandez-Lizarbe S, Pascual M, Gascon MS, Blanco A, Guerri C. Lipid rafts regulate ethanol-induced activation of TLR4 signaling in murine macrophages. Mol Immunol 2008; 45(7): 2007-16.
[http://dx.doi.org/10.1016/j.molimm.2007.10.025] [PMID: 18061674]

[40] Montesinos J, Pascual M, Pla A, *et al.* TLR4 elimination prevents synaptic and myelin alterations and long-term cognitive dysfunctions in adolescent mice with intermittent ethanol treatment. Brain Behav Immun 2015; 45: 233-44.

[http://dx.doi.org/10.1016/j.bbi.2014.11.015] [PMID: 25486089]

[41]   Montesinos J, Pascual M, Rodríguez-Arias M, Miñarro J, Guerri C. Involvement of TLR4 in the long-term epigenetic changes, rewarding and anxiety effects induced by intermittent ethanol treatment in adolescence. Brain Behav Immun 2016; 53: 159-71.
[http://dx.doi.org/10.1016/j.bbi.2015.12.006] [PMID: 26686767]

[42]   Montesinos J, Pascual M, Millán-Esteban D, Guerri C. Binge-like ethanol treatment in adolescence impairs autophagy and hinders synaptic maturation: Role of TLR4. Neurosci Lett 2018; 682: 85-91.
[http://dx.doi.org/10.1016/j.neulet.2018.05.049] [PMID: 29864452]

[43]   Blanco AM, Guerri C. Ethanol intake enhances inflammatory mediators in brain: Role of glial cells and TLR4/IL-1RI receptors. Front Biosci 2007; 12(1): 2616-30.
[http://dx.doi.org/10.2741/2259] [PMID: 17127267]

[44]   Blanco AM, Vallés SL, Pascual M, Guerri C. Involvement of TLR4/type I IL-1 receptor signaling in the induction of inflammatory mediators and cell death induced by ethanol in cultured astrocytes. J Immunol 2005; 175(10): 6893-9.
[http://dx.doi.org/10.4049/jimmunol.175.10.6893] [PMID: 16272348]

[45]   Blanco AM, Perez-Arago A, Fernandez-Lizarbe S, Guerri C. Ethanol mimics ligand-mediated activation and endocytosis of IL-1RI/TLR4 receptors *via lipid rafts* caveolae in astroglial cells. J Neurochem 2008; 106(2): 625-39.
[http://dx.doi.org/10.1111/j.1471-4159.2008.05425.x] [PMID: 18419766]

[46]   Pascual-Lucas M, Fernandez-Lizarbe S, Montesinos J, Guerri C. LPS or ethanol triggers clathrin- and rafts/caveolae-dependent endocytosis of TLR 4 in cortical astrocytes. J Neurochem 2014; 129(3): 448-62.
[http://dx.doi.org/10.1111/jnc.12639] [PMID: 24345077]

[47]   Pascual M, Pla A, Miñarro J, Guerri C. Neuroimmune activation and myelin changes in adolescent rats exposed to high-dose alcohol and associated cognitive dysfunction: A review with reference to human adolescent drinking. Alcohol Alcohol 2014; 49(2): 187-92.
[http://dx.doi.org/10.1093/alcalc/agt164] [PMID: 24217958]

[48]   Li Q, Liu D, Pan F, Ho CSH, Ho RCM. Ethanol exposure induces microglia activation and neuroinflammation through TLR4 activation and SENP6 modulation in the adolescent rat hippocampus. Neural Plast 2019; 2019: 1-12.
[http://dx.doi.org/10.1155/2019/1648736] [PMID: 31781182]

[49]   Pla A, Pascual M, Renau-Piqueras J, Guerri C. TLR4 mediates the impairment of ubiquitin-proteasome and autophagy-lysosome pathways induced by ethanol treatment in brain. Cell Death Dis 2014; 5(2): e1066.
[http://dx.doi.org/10.1038/cddis.2014.46] [PMID: 24556681]

[50]   Ibáñez F, Montesinos J, Ureña-Peralta JR, Guerri C, Pascual M. TLR4 participates in the transmission of ethanol-induced neuroinflammation *via* astrocyte-derived extracellular vesicles. J Neuroinflammation 2019; 16(1): 136.
[http://dx.doi.org/10.1186/s12974-019-1529-x] [PMID: 31272469]

[51]   Alfonso-Loeches S, Ureña-Peralta J, Morillo-Bargues MJ, Gómez-Pinedo U, Guerri C. Ethanol-induced TLR4/NLRP3 neuroinflammatory response in microglial cells promotes leukocyte infiltration across the BBB. Neurochem Res 2016; 41(1-2): 193-209.
[http://dx.doi.org/10.1007/s11064-015-1760-5] [PMID: 26555554]

[52]   Tajuddin N, Moon KH, Marshall SA, *et al.* Neuroinflammation and neurodegeneration in adult rat brain from binge ethanol exposure: Abrogation by docosahexaenoic acid. PLoS One 2014; 9(7): e101223.
[http://dx.doi.org/10.1371/journal.pone.0101223] [PMID: 25029343]

[53]   Tajuddin NF, Przybycien-Szymanska MM, Pak TR, Neafsey EJ, Collins MA. Effect of repetitive daily ethanol intoxication on adult rat brain: Significant changes in phospholipase A2 enzyme levels in

association with increased PARP-1 indicate neuroinflammatory pathway activation. Alcohol 2013; 47(1): 39-45.
[http://dx.doi.org/10.1016/j.alcohol.2012.09.003] [PMID: 23102656]

[54]   Moon KH, Tajuddin N, Brown J III, Neafsey EJ, Kim HY, Collins MA. Phospholipase A2, oxidative stress, and neurodegeneration in binge ethanol-treated organotypic slice cultures of developing rat brain. Alcohol Clin Exp Res 2014; 38(1): 161-9.
[http://dx.doi.org/10.1111/acer.12221] [PMID: 23909864]

[55]   Tajuddin N, Kim HY, Collins MA. PARP inhibition prevents ethanol-induced neuroinflammatory signaling and neurodegeneration in rat adult-age brain slice cultures. J Pharmacol Exp Ther 2018; 365(1): 117-26.
[http://dx.doi.org/10.1124/jpet.117.245290] [PMID: 29339456]

[56]   Barnett A, David E, Rohlman A, *et al.* Adolescent binge alcohol enhances early alzheimer's disease pathology in adulthood through proinflammatory neuroimmune activation. Front Pharmacol 2022; 13: 884170.
[http://dx.doi.org/10.3389/fphar.2022.884170] [PMID: 35559229]

[57]   Tucker AE, Alicea Pauneto CM, Barnett AM, Coleman LG Jr. Chronic ethanol causes persistent increases in alzheimer's tau pathology in female 3xTg-AD mice: A potential role for lysosomal impairment. Front Behav Neurosci 2022; 16: 886634.
[http://dx.doi.org/10.3389/fnbeh.2022.886634] [PMID: 35645744]

[58]   Hoffman JL, Faccidomo S, Kim M, *et al.* Alcohol drinking exacerbates neural and behavioral pathology in the 3xTg-AD mouse model of Alzheimer's disease. Int Rev Neurobiol 2019; 148: 169-230.
[http://dx.doi.org/10.1016/bs.irn.2019.10.017] [PMID: 31733664]

[59]   Vetreno RP, Lawrimore CJ, Rowsey PJ, Crews FT. Persistent adult neuroimmune activation and loss of hippocampal neurogenesis following adolescent ethanol exposure: Blockade by exercise and the anti-inflammatory drug indomethacin. Front Neurosci 2018; 12: 200.
[http://dx.doi.org/10.3389/fnins.2018.00200] [PMID: 29643762]

[60]   Vetreno RP, Crews FT. Binge ethanol exposure during adolescence leads to a persistent loss of neurogenesis in the dorsal and ventral hippocampus that is associated with impaired adult cognitive functioning. Front Neurosci 2015; 9: 35.
[http://dx.doi.org/10.3389/fnins.2015.00035] [PMID: 25729346]

[61]   Macht V, Vetreno R, Elchert N, Crews F. Galantamine prevents and reverses neuroimmune induction and loss of adult hippocampal neurogenesis following adolescent alcohol exposure. J Neuroinflammation 2021; 18(1): 212.
[http://dx.doi.org/10.1186/s12974-021-02243-7] [PMID: 34530858]

[62]   Macht V, Crews FT, Vetreno RP. Neuroimmune and epigenetic mechanisms underlying persistent loss of hippocampal neurogenesis following adolescent intermittent ethanol exposure. Curr Opin Pharmacol 2020; 50: 9-16.
[http://dx.doi.org/10.1016/j.coph.2019.10.007] [PMID: 31778865]

[63]   Sakharkar AJ, Vetreno RP, Zhang H, Kokare DM, Crews FT, Pandey SC. A role for histone acetylation mechanisms in adolescent alcohol exposure-induced deficits in hippocampal brain-derived neurotrophic factor expression and neurogenesis markers in adulthood. Brain Struct Funct 2016; 221(9): 4691-703.
[http://dx.doi.org/10.1007/s00429-016-1196-y] [PMID: 26941165]

[64]   Kyzar EJ, Zhang H, Sakharkar AJ, Pandey SC. Adolescent alcohol exposure alters lysine demethylase 1 (LSD1) expression and histone methylation in the amygdala during adulthood. Addict Biol 2017; 22(5): 1191-204.
[http://dx.doi.org/10.1111/adb.12404] [PMID: 27183824]

[65]   Sakharkar AJ, Kyzar EJ, Gavin DP, *et al.* Altered amygdala DNA methylation mechanisms after

adolescent alcohol exposure contribute to adult anxiety and alcohol drinking. Neuropharmacology 2019; 157: 107679.
[http://dx.doi.org/10.1016/j.neuropharm.2019.107679] [PMID: 31229451]

[66]   Crews FT, Robinson DL, Chandler LJ, *et al.* Mechanisms of persistent neurobiological changes following adolescent alcohol exposure: NADIA consortium findings. Alcohol Clin Exp Res 2019; 43(9): 1806-22.
[http://dx.doi.org/10.1111/acer.14154] [PMID: 31335972]

[67]   Zou J, Walter TJ, Barnett A, Rohlman A, Crews FT, Coleman LG Jr. Ethanol induces secretion of proinflammatory extracellular vesicles that inhibit adult hippocampal neurogenesis through G9a/GLP-epigenetic signaling. Front Immunol 2022; 13: 866073.
[http://dx.doi.org/10.3389/fimmu.2022.866073] [PMID: 35634322]

[68]   Khatri D, Laroche G, Grant ML, *et al.* Acute ethanol inhibition of adult hippocampal neurogenesis involves CB 1 cannabinoid receptor signaling. Alcohol Clin Exp Res 2018; 42(4): 718-26.
[http://dx.doi.org/10.1111/acer.13608] [PMID: 29417597]

[69]   Broadwater MA, Liu W, Crews FT, Spear LP. Persistent loss of hippocampal neurogenesis and increased cell death following adolescent, but not adult, chronic ethanol exposure. Dev Neurosci 2014; 36(3-4): 297-305.
[http://dx.doi.org/10.1159/000362874] [PMID: 24993092]

[70]   Liu W, Crews FT. Persistent decreases in adult subventricular and hippocampal neurogenesis following adolescent intermittent ethanol exposure. Front Behav Neurosci 2017; 11: 151.
[http://dx.doi.org/10.3389/fnbeh.2017.00151] [PMID: 28855864]

[71]   Kokare DM, Kyzar EJ, Zhang H, Sakharkar AJ, Pandey SC. Adolescent alcohol exposure-induced changes in alpha-melanocyte stimulating hormone and neuropeptide Y pathways *via* histone acetylation in the brain during adulthood. Int J Neuropsychopharmacol 2017; 20(9): 758-68.
[http://dx.doi.org/10.1093/ijnp/pyx041] [PMID: 28575455]

[72]   Macht VA, Vetreno RP, Crews FT. Cholinergic and neuroimmune signaling interact to impact adult hippocampal neurogenesis and alcohol pathology across development. Front Pharmacol 2022; 13: 849997.
[http://dx.doi.org/10.3389/fphar.2022.849997] [PMID: 35308225]

[73]   Vetreno RP, Crews FT. Adolescent binge ethanol-induced loss of basal forebrain cholinergic neurons and neuroimmune activation are prevented by exercise and indomethacin. PLoS One 2018; 13(10): e0204500.
[http://dx.doi.org/10.1371/journal.pone.0204500] [PMID: 30296276]

Nicotine

Abstract: Nicotine is present in the tobacco products. Once smoked, nicotine immediately reaches the brain and binds with nicotinic receptors causing damage to the brain cells. The adolescent brain is especially very sensitive to products such as e-cigarettes, nicotine, and tobacco. Chronic nicotine exposure causes permanent brain damage and cognitive decline. Interestingly there are reports on the use of nicotine and its effects on the epigenetic changes in the brain. These kinds of changes may prepare the brain for further abuse of various illegal drugs. As a result of chronic nicotine exposure brain infarcts, white matter hyperintensities, brain atrophy, and dementia are also known to occur. Neurodevelopment in children is potentially harmed due to exposure to nicotine and nicotinic products. This is due to the inflammation, atherosclerosis, and oxidative stress to the neuronal cells. Pregnant mothers and people who are at risk of developing neurodegenerative disease need to be forbidden from using nicotine. Nicotine can be dangerous when taken with alcohol as it can lead to depression and neurocognitive decline. This chapter addresses the effects of nicotine on the adolescent and adult brain.

Keywords: Brain atrophy, Cognitive decline, Depression, Neurodevelopment, Oxidative stress.

INTRODUCTION

Nicotine is a product from tobacco that is consumed as different products such as E-cigarettes and smoking. Nicotine is a plant-based alkaloid that binds with the acetylcholine receptors in the brain and mimics the effects of acetylcholine. Not only CNS disorders, smoking tobacco by various methods causes various diseases such as cancer, cardiovascular diseases, and respiratory diseases. Nicotine not only influences the acetylcholine receptors in the brain but also has an impact on dopamine, serotonin, and norepinephrine. When these neurotransmitters are affected it leads to cognitive, motor, learning, and memory dysfunctions, especially the executive functions. The reinforcement of smoking initially is due to the enhancement of cognitive enhancement especially attention and memory, however, at later stages, it leads to cognitive impairment and cognitive decline [1].

Nicotine in the Ageing Brain

Brain ageing is a very complicated process. The ageing brain displays various changes in its morphology, biochemistry, and physiology. Brain ageing due to reductions in acetylcholinergic pathways (degradation of neurons in the nucleus of myenert) may lead to the development of oxidative stress, beta-amyloid toxicity, calcium-related singlaing dysfunctions, reduced neurotrophic factors, neuroinflammation, and apoptosis. In this condition, smoking by old age persons will lead to the destruction of molecular pathways and dysregulation of nicotinic Acetylcholinergic pathways [2]. However, some studies on animals suggest that nicotine can be neuroprotective instead of neuro-damaging. Activation of nicotinic acetylcholine receptors can improve memory and learning [3]. Controversially, some studies have shown that smoking helps a little delay in developing neurodegeneration.

Nicotine and Fetal Brain Development

Smoking during pregnancy can lead to devastating effects on the developing fetus. Some fetal neurodevelopmental disorders are due to the nicotine smoking by pregnant mothers. The US Food and Drug Administration agency has classified nicotine as a class drug during pregnancy. Pregnancy-related morbidity and mortality is most of the time due to smoking cigarettes or tobacco products. Nicotine leads to reduced blood flow to the placenta leading to cardiovascular dysfunctions, which has been proved by animal studies. Smoking replacement therapy (NRT) during pregnancy has not been well studied [4].

Nicotine and Adolescent Brain

Inhalation of tobacco products can cause other comorbid situations such as anxiety disorders, mood disorders, and Schizophrenia. People with these disorders when consuming tobacco lead to various modulations in the brain circuits. Nicotine influences various neurotransmitters such as GABA, Glutamate, and Dopamine all these neurotransmitters are implicated in various neurodegenerative disorders. Largely, it has been observed that many neurological disorders such as neuropsychiatric disorders are developed either during the developmental period or during the adolescent time of development [5]. The mesocortical limbic system is influenced by mood and anxiety-related symptoms due to exposure to nicotine during adolescence time. The studies performed in rats during adolescence time confer the long-lasting changes in the mesolimbic system and involve disturbances in PFC DA D1R and downstream extracellular-signal-related kinase 1-2 (ERK 1-2) pathways [6]. The nucleus accumbent shell is important for emotional processing. Any disturbances in this structure will lead to the development of mood and anxiety-related disorders. Using an adolescent rat model several molecular markers which are related to nicotine exposure were examined. ERK 1-2 and Akt-GSK-3 levels were much higher in the adolescent brain but not in the adult brain [7]. Further studies on rats reconfirm that maternal deprivation stress induced the rats' drug-seeking behaviour affecting the amygdala and ventral tegmental area in the brain [8]. Cognitive and emotional disturbances were noted in rodents exposed to prenatal nicotine exposure thus in turn affecting the glutamate receptor-associated genes in the prefrontal cortex [9]. Nicotine abuse in the brain is attributed to the mesolimbic dopamine system, thus affecting the reward stimuli and sensory systems in the early stages of addiction.

Nicotinic Acetylcholine Receptors (nAChRs) in Addiction to Tobacco Smoking

Nicotine's ability to enhance dopamine firing is due to its binding to various subtypes of nicotinic receptors such as alpha 6 [10]. Nicotinic acetylcholine receptors (nAChRs) are ligand-gated ion channels spanning across the membrane with five subunits. α4 and β2 subunits are considered as the predominant subunits of the nAChRs.Almost all subunits of the nAChRs are involved in addiction due to tobacco smoking. α4β2* nAChRs i the main therapeutic target for the approved drugs against the cessation of smoking [11]. There are twelve α subunits (α2-α10) and three beta subunits (β2-β4) of nAChRs in mammalian brain. The receptors such as α4β2 nAChRs, play a very important role in reinforcing qualities of the nicotine, particularly in the mesoaccumbens dopamine pathway. Cloning studies have revealed the genes located in chromosome 15q25, which encode the α5, α3, and β4 nAChR subunits show a particular pattern of genetic variation.

CONCLUSION

Nicotine is a product from tobacco that is consumed as different products such as E-cigarettes and smoking. Nicotine is a plant-based alkaloid that binds with the acetylcholine receptors in the brain and mimics the effects of acetylcholine. Not only CNS disorders, smoking tobacco by various methods causes various diseases such as cancer, cardiovascular diseases, and respiratory diseases. Nicotine not only influences the acetylcholine receptors in the brain but also has an impact on dopamine, serotonin, and norepinephrine. Brain aging is a very complicated process. The aging brain displays various changes in its morphology, biochemistry, and physiology. Brain ageing due to reductions in acetylcholinergic pathways (degradation of neurons in the nucleus of myenert) may lead to the development of oxidative stress, beta-amyloid toxicity, calcium-related singlaing dysfunctions, reduced neurotrophic factors, neuroinflammation, and apoptosis. In this condition, smoking by old age persons will lead to the destruction of molecular pathways and dysregulation of nicotinic Acetylcholinergic pathways.

Smoking during pregnancy can lead to devastating effects on the developing fetus. Some fetal neurodevelopmental disorders are due to nicotine smoking by pregnant mothers.

REFERENCES

[1]   Campos MW, Serebrisky D, Castaldelli-Maia JM. Smoking and cognition. Curr Drug Abuse Rev 2017; 9(2): 76-9.
[http://dx.doi.org/10.2174/1874473709666160803101633] [PMID: 27492358]

[2]   Majdi A, Kamari F, Vafaee MS, Sadigh-Eteghad S. Revisiting nicotine's role in the ageing brain and cognitive impairment. Rev Neurosci 2017; 28(7): 767-81.
[http://dx.doi.org/10.1515/revneuro-2017-0008] [PMID: 28586306]

[3]   Üzüm G, Díler AS, Bahçekapili N, Tasyüreklí M, Zíylan Z. Nicotine improves learning and memory in rats: Morphological evidence for acetylcholine involvement. Int J Neurosci 2004; 114(9): 1163-79.
[http://dx.doi.org/10.1080/00207450490475652] [PMID: 15370181]

[4]   Dempsey DA, Benowitz NL. Risks and benefits of nicotine to aid smoking cessation in pregnancy. Drug Saf 2001; 24(4): 277-322.
[http://dx.doi.org/10.2165/00002018-200124040-00005] [PMID: 11330657]

[5]   Laviolette SR. Molecular and neuronal mechanisms underlying the effects of adolescent nicotine exposure on anxiety and mood disorders. Neuropharmacology 2021; 184: 108411.
[http://dx.doi.org/10.1016/j.neuropharm.2020.108411] [PMID: 33245960]

[6]   Jobson CLM, Renard J, Szkudlarek H, *et al.* Adolescent nicotine exposure induces dysregulation of mesocorticolimbic activity states and depressive and anxiety-like prefrontal cortical molecular phenotypes persisting into adulthood. Cereb Cortex 2019; 29(7): 3140-53.
[http://dx.doi.org/10.1093/cercor/bhy179] [PMID: 30124787]

[7]   Hudson R, Green M, Wright DJ, *et al.* Adolescent nicotine induces depressive and anxiogenic effects through ERK 1-2 and Akt-GSK-3 pathways and neuronal dysregulation in the nucleus accumbens. Addict Biol 2021; 26(2): e12891.
[http://dx.doi.org/10.1111/adb.12891] [PMID: 32135573]

[8]   Bassey RB, Gondré-Lewis MC. Combined early life stressors: Prenatal nicotine and maternal deprivation interact to influence affective and drug seeking behavioral phenotypes in rats. Behav Brain Res 2019; 359: 814-22.
[http://dx.doi.org/10.1016/j.bbr.2018.07.022] [PMID: 30055209]

[9]   Polli FS, Scharff MB, Ipsen TH, Aznar S, Kohlmeier KA, Andreasen JT. Prenatal nicotine exposure in mice induces sex-dependent anxiety-like behavior, cognitive deficits, hyperactivity, and changes in the expression of glutamate receptor associated-genes in the prefrontal cortex. Pharmacol Biochem Behav 2020; 195: 172951.
[http://dx.doi.org/10.1016/j.pbb.2020.172951] [PMID: 32439454]

[10]   Wittenberg RE, Wolfman SL, De Biasi M, Dani JA. Nicotinic acetylcholine receptors and nicotine addiction: A brief introduction. Neuropharmacology 2020; 177: 108256.
[http://dx.doi.org/10.1016/j.neuropharm.2020.108256] [PMID: 32738308]

[11]   Picciotto MR, Kenny PJ. Molecular mechanisms underlying behaviors related to nicotine addiction. Cold Spring Harb Perspect Med 2013; 3(1): a012112.
[http://dx.doi.org/10.1101/cshperspect.a012112] [PMID: 23143843]

SUBJECT INDEX

www.ingramcontent.com/pod-product-compliance
Lightning Source LLC
Chambersburg PA
CBHW041449210326
41599CB00004B/190